MW01252669

First Responders Handbook

Bob
Congratulations & Best Wishes
for all your accomplishments
and we hope that you
enjoy your retirement.
Love
Mom & Dad

LTC Blonigen

Congratulations on your retirement, we are All soldiers no matter what service we served in, and thank you for your service and sacrifices to our country.

Only those of us that have served know the true meaning and understanding of what being an American is all about.

Placing all other before ourselves, to accomplish what is just and right, and knowing we have made this world a better place live.

In the end it's all about - Duty, Honor, and Country

God Bless, and Thank you for your service

Soldier to Soldier

Michael Madigan
SFC, USA
Retired

First Responders Handbook

An Introduction

Second Edition

By
Michael L. Madigan

CRC Press is an imprint of the
Taylor & Francis Group, an **informa** business

CRC Press
Taylor & Francis Group
6000 Broken Sound Parkway NW, Suite 300
Boca Raton, FL 33487-2742

Printed on acid-free paper

International Standard Book Number-13: 978-1-138-08968-6 (Hardback)

Library of Congress Cataloging-in-Publication Data

Names: Madigan, Michael L., author.
Title: First responders handbook: An introduction / Michael L. Madigan.
Description: Second edition. | Boca Raton : Taylor & Francis, 2017. | "A CRC title, part of the Taylor & Francis imprint, a member of the Taylor & Francis Group, the academic division of T&F Informa plc." | Includes bibliographical references.
Identifiers: LCCN 2017012370| ISBN 9781138089686 (hardback : alk. paper) | ISBN 9781315109114 (ebook)
Subjects: LCSH: Emergency medical services--Handbooks, manuals, etc. | First responders--Handbooks, manuals, etc.
Classification: LCC RA645.5 .M33 2017 | DDC 362.18--dc23
LC record available at https://lccn.loc.gov/2017012370

Visit the Taylor & Francis Web site at
http://www.taylorandfrancis.com

and the CRC Press Web site at
http://www.crcpress.com

Contents

Preface xvii
About the Author xix
Introduction xxi

Chapter 1 NIMS and the incident command system 1
Introduction 1
The history of the incident command system 2
National Incident Management System 3
What is NIMS ICS? 4
ICS management 4
Available NIMS ICS training 6
Conclusion 10

Chapter 2 Security management 11
Introduction 11
Types of security threats 11
External 11
Internal 12
Risk 12
Risk avoidance 12
Risk reduction 12
Risk spreading 12
Risk transfer 12
Risk acceptance 13
Security policy implementations 13
Intrusion detection 13
Access control 14
Physical security 15
Law enforcement agency (LEA) jurisdiction procedures 16
Non-executive powers jurisdictional coverage of Europol 16
Fraud management 17
Alarm management 17
Concepts 17

The need for alarm management ... 19
Some improvement methods ... 19
Nuisance reduction ... 19
Design guide ... 19
Documentation and rationalization ... 19
Advanced methods ... 20
The seven steps to alarm management ... 20
Step 1: Create and adopt an alarm philosophy ... 20
Step 2: Alarm performance benchmarking ... 20
Step 3: "Bad actor" alarm resolution ... 20
Step 4: Alarm documentation and rationalization (D&R) ... 20
Step 5: Alarm system audit and enforcement ... 21
Step 6: Real-time alarm management ... 21
Step 7: Control and maintain alarm system performance ... 21
Information technology (IT) risk ... 21
Committee on National Security Systems ... 21
Measuring IT risk ... 22
Occurrence of a particular set of circumstances ... 22
Estimation of likelihood ... 23
Estimation of technical impact ... 24
Estimation of business impact ... 25
Risk management as part of enterprise risk management ... 27
Risk-management methodology ... 28
Perceived security compared to real security ... 28
Categorizing security ... 29
Security concepts ... 30
Significance ... 30
Conclusion ... 31

Chapter 3 Threat/vulnerability assessments and risk analysis ... 33
Introduction ... 33
Threat assessment ... 34
Vulnerability assessment ... 35
Risk analysis ... 37
Upgrade recommendations ... 37
Re-evaluation of risks ... 39
Understanding threats ... 39
Electronic risk assessment and vulnerabilities: Computer/network internal and external threats ... 43
IT security risk assessment and analysis: Reporting objectives ... 44
Methodology ... 44
The threat vulnerability assessment tool ... 44
Risk analysis description ... 44
Threat and vulnerability analysis ... 45

Threat and vulnerability analysis description 45
Attacks and threats 45
Principles of conducting risk characterizations 46
Conclusion 48

Chapter 4 Defense support of civil authorities (DSCA) 49
Defense support of civil authorities 49
MOOTW also involves arms control and peacekeeping 52
Select American deployments 53
Selected British deployments 53
Select Japanese deployments 54
Select Australian deployments 54
Current Australian deployments 54
Conclusion 54

Chapter 5 Introduction to weapons of mass destruction (WMDs) 61
Chemical weapons of mass destruction 62
Sources of CW agents 62
Risks from chemical agents 62
Recent chemical terrorism events 62
Classification of chemical weapons 62
Chief categories of agents 63
Nerve agents 63
Vesicants/blister agents 63
Pulmonary damaging agents 64
Blood agents or cyanides 64
Riot-control agents 64
General treatment guidelines for all classes of chemical weapons 65
Biological weapons of mass destruction 65
What is bioterrorism? 65
Why biologics are attractive to terrorists 65
Characteristics of biological attacks 65
Genealogical classification of bioterrorism agents 65
CDC: Critical biological agents 66
Category A 66
Category B 66
Category C 66
Category A bioterrorism agents 66
Category B bioterrorism agents 67
Category C bioterrorism agents 67
Smallpox 67
History of smallpox 67
Variola major (smallpox) 67
Smallpox: Clinical features 68

Smallpox versus chickenpox.....68
Smallpox vaccine.....68
Anthrax.....68
Overview.....68
Anthrax: Cutaneous.....69
Anthrax: Gastrointestinal.....69
Anthrax: Inhalational.....69
Anthrax: Vaccine.....69
Plague.....70
Overview.....70
Plague: Bubonic.....70
Plague: Pneumonic.....70
Plague: Septicemic.....70
Radioactive and nuclear weapons of mass destruction.....71
Radiation versus radioactive material.....71
Exposure versus contamination.....71
Injuries associated with radiation exposure.....72
Acute radiation syndrome.....72
Cutaneous radiation syndrome.....72
Detecting radiation: Survey meters.....72
Detecting radiation: Dosimeter devices.....73
Methods of protection.....73
Potassium iodide (KI) tablets.....73
Radioactive/nuclear WMDs: Possible scenarios.....73
Nuclear power plant incident.....73
Nuclear weapon.....73
Improvised nuclear device.....73
"Dirty bomb".....74
Explosive/incendiary weapons of mass destruction.....74
The most widely used WMDs!.....74
Toxic industrial chemicals and toxic industrial materials.....74
RAIN.....75
Recognition.....75
Toxic industrial chemicals/materials (TIC/TIM).....75
Using the Emergency Response Guidebook (ERG) 2012.....75
Sources of toxic industrial chemicals/toxic industrial materials.....76
Chemical warfare agents (using industrial chemicals).....77
Choking agents.....77
Blood agents (Cyanides).....77
Blister agents.....78
Nerve agents.....78
Lethality relative to chlorine.....78
Isolate.....79
Control Zones.....79

Advantages/disadvantages of using chemical agents as WMDs......... 80
Advantages 80
Disadvantages 81
Cyberwarfare 81
Espionage and national security breaches 81
Sabotage 81
Denial-of-service attack 82
Electrical power grid 82
Military 83
Terrorism 83
Civil 84
Private sector 84
Non-profit research 84
National Incident Management System/Incident Command System (NIMS/ICS) 84
National Integration Center (NIC) guidance delivery and whole-community engagement 85
National Integration Center (NIC) 86
Incident Command System 86
ICS organization 87
Modular expansion 87
Command staff 89
Command functions/sections 89
Establishing branches to maintain recommended span of control 90
First responder categories and capabilities 90
First responder awareness level guidelines 90
Law enforcement 90
Fire fighters 91
Emergency medical service (EMS) providers 91
Emergency management personnel 92
Public works employees 92
First responders' performance level guidelines 93
Law enforcement 93
Fire service 94
EMS 95
Hazardous materials 96
Public works 97
Planning and management level guidelines 98
Law enforcement 98
Fire service 99
EMS 100
HAZMAT 101
Emergency management 102
Public works 103

First Responders' Resources....104
WMD civil support team (WMD-CST)....104
CBRNE enhanced response force package (CERFP)....106
Conclusion....106

Chapter 6 Understanding terror, terrorism, and their roots....107
What is terrorism?....108
Why talk about terrorism?....109
Official U.S. government definition of terrorism....110
General history of terrorism....111
The origins of modern terrorism....111
The rise of non-state terrorism....111
Terrorism turns international....112
The twenty-first century: Religious terrorism and beyond....112
Violence in the Koran and the Bible....112
The teaching of Islam about Jihad or fighting for the cause of Allah...113
What is Islam?....113
Islamic radicals....114
Terrorist tactics....117
Improvised explosive devices....117
The internet and psychological warfare....117
Terrorist training camps....118
Terrorist incident activities....118
The cost of terrorism....119
Conclusion....121

Chapter 7 CBRNE and weapons of mass destruction....123
Chemical weapons (CW)....123
Biological weapons (BW)....126
Pathogen feed stocks....127
BW delivery systems....128
Radiological weapons (RW)....128
Nuclear weapons (NW)....129
Industrial chemicals as weapons of mass destruction....131
Tactics....131
Industrial chemicals....132
Industrial chemicals that can be weaponized....133
Chemical agents and terrorism....134
Indicators of a possible chemical incident....134
Conclusion....135

Chapter 8 Explosives....137
Chemical....138
Decomposition....138

Deflagration 138
Detonation 138
Exotic 138
Properties of explosive materials 139
Availability and cost 139
Sensitivity 139
Sensitivity to initiation 140
Velocity of detonation 140
Stability 140
Power, performance, and strength 141
Brisance 142
Density 143
Volatility 143
Hygroscopicity and water resistance 143
Toxicity 144
Explosive train 144
Volume of products of explosion 144
Oxygen balance (OB% or Ω) 145
Chemical composition 145
Chemically pure compounds 145
Mixture of oxidizer and fuel 146
Classification of explosive materials 146
By sensitivity 146
Primary explosive 146
Secondary explosive 147
Tertiary explosive 147
By velocity 147
Low explosives 147
High explosives 150
By composition 150
Priming composition 150
By physical form 150
Shipping label classifications 150
United Nations Organization (UNO) Hazard Class and Division (HC/D) 151
Explosives warning sign 151
Class 1 Compatibility Group 151
Commercial application 153
Explosive velocity 154
Natural 155
Astronomical 155
Chemical 156
Electrical and magnetic 156
Mechanical and vapor 156

Nuclear 156
Properties of explosions 157
Force 157
Velocity 157
Evolution of heat 157
Initiation of reaction 158
Fragmentation 158
Conclusion 159

Chapter 9 Nuclear weapons and radiation 161
A brief history of nuclear weapons 161
Fission weapons 162
Fusion weapons 163
Thermonuclear weapons 163
Nuclear weapons: Other uses 164
Nuclear weapons delivery systems 165
Nuclear strategy 166
Nuclear terrorism 168
Militant groups 169
Incidents involving nuclear material 170
Conclusion 171

Chapter 10 Dirty bomb 175
Impact of a dirty bomb 175
Protective actions 176
Sources of radioactive material 176
Control of radioactive material 176
Risk of cancer 177
Other contact information 177
What is an RDD or "dirty bomb"? 178
What is radiation? 178
Are terrorists interested in radioactive materials? 178
Will an RDD make me sick? 179
How can I protect myself in a radiation emergency? 179
Is it necessary to purchase potassium iodide tablets for protection against radiation? 180
Conclusion 180

Chapter 11 Decontamination procedures 181
Definition of decontamination 181
Prevention of contamination 181
Types of contamination 182
Decontamination methods 182

Testing the effectiveness of decontamination 183
Decontamination plan 184
Decontamination for protective clothing reuse 185
Emergency decontamination 185
Conclusion 186

Chapter 12 Decontamination of chemical warfare and industrial agents 187
Introduction to decontamination methods and procedures 187
Decontaminants 187
Decontamination methods 188
Individual decontamination 189
Decontamination of equipment 190
Guidelines for mass casualty decontamination during a terrorist chemical agent incident 192
Relationships and process 192
Purposes of decontamination 193
Methods of mass decontamination 194
Decontamination procedures 195
Decontamination approaches 196
Field-expedient water decontamination methods 196
Types of chemical victims 197
Prioritizing casualties for decontamination 198
Triage definitions 198
Mass casualty/casualty decontamination processing tag system 200
Casualty processing 200
Additional considerations 201
Environmental concerns 201
Conclusion 203

Chapter 13 Chemical protective clothing 205
Protective clothing applications 205
Level of protection 207
Ensemble selection factors 207
Classification of protective clothing 209
Service life of protective clothing 213
Protective clothing selection factors 214
Sources of information 215
Physical properties 215
General guidelines 216
Clothing donning, doffing, and use 218
Donning the ensemble 218
Doffing an ensemble 220
User monitoring and training 221

Inspection, storage, and maintenance 221
Inspection 222
Storage 222
Maintenance 223
Training 224
Risks 224
Heat stress 224
Heart rate 225
Oral temperature 225
Body water loss 225
Conclusion 225

Chapter 14 Emergency response guidebook 227
Guidebook contents 227
White section (front) 228
Yellow section 229
Blue section 229
Orange section 229
Green section 229
White section (back) 230
Shipping list 230
Contents 230
Recipients 230
Shipping list details 231
Classification and labeling summary tables 231
Class 1: Explosives 231
Class 2: Gases 232
Class 3: Flammable liquids 232
Class 4: Flammable solids 232
Class 5: Oxidizing agents and organic peroxides 233
Class 6: Toxic and infectious substances 233
Class 7: Radioactive substances 233
Class 8: Corrosive substances 233
Class 9: Miscellaneous 233
Boiling liquid expanding vapor explosion (BLEVE) 233
Mechanism 233
Water example 234
BLEVEs without chemical reactions 235
Fires 235
Globally Harmonized System of Classification and Labeling of Chemicals 236
Hazard classification 236
Physical hazards 236
Health hazards 238

Environmental hazards 240
Classification of mixtures 240
Hazard communication 240
GHS label elements 241
GHS label format 242
GHS material safety data sheet or safety data sheet 243
Training 244
Implementation 244
Conclusion 244

Glossary 247
Bibliography 263
Index 267

Preface

This book provides a general approach to natural disasters and terrorist or criminal incidents from the aspect of the first responder community. The book examines human elements, such as the radical fundamentalist terrorist, as well as the security and policies of the United States, how and why we respond as a nation, and how we respond on the local, state, and federal level.

The above mentions the human element, but too many times we forget how to perform and react in these situations, the step-by-step procedures for how we as nation respond, and how first responders should act and behave throughout the response.

I am focusing on practical applications, from real-world experiences of my own and the actions of other first responders that I have come in contact with over the years. These responders come from the Department of Defense, law enforcement, fire services, and the emergency medical services throughout the United States, and our allies from sister elements of law enforcement, fire, and emergency medical services, and other military respond forces.

About the Author

Michael L. Madigan works as a master instructor for the Civil Support Skills Course (CSTs) teams, and as a specialist in weapons of mass destruction hazardous material emergency response operations. He delivers threat assessments to the Homeland Defense Civil Support Office (HD/CSO), and the Incident Response Training Department (IRTD) on Fort Leonard Wood in Missouri. He is also a certified FEMA instructor for ICS courses 300 and ICS 400. He is retired from the United States Army, having served as a (SME), master chemical operation specialist to the USACBRN School, CENTCOM, and the Multi-National Forces Command in IRAQ.

He holds a BA degree in Criminal Justice from the University of Massachusetts Lowell, and a Master of Security Management from the American Military University at Charles Town. In addition, he holds the Certificate in Security Management from the University of Massachusetts.

Over my 40-year career, I have distinguished myself by exceptionally meritorious service, in a succession of positions of greater importance and responsibility, to the Army and the Nation, culminating in the Strategy Operations Plans and Policy Directorate (G-3/5/7), Headquarters, Department of the Army. My previous positions of significant leadership included Operations J3 Multi National Forces Iraq (MNFI), and WMD/Terrorism/CBRNE NCOIC Team Chief.

As the WMD/Terrorism and CBRNE Team Chief I have positively and directly influenced Army and Department of Defense Policy and program recommendations to the MNFI Commander. I have carefully communicated advice reflecting the depth and breadth of my considerable experience, and affected future programs and policy that will carry the Army through the end of sustained combat operations and into a future focused on building the capacity of the partners of the United States.

As the NCOIC of the WMD/Terrorism/CBRNE, I played a key role in the transition between Operation Iraqi Freedom and Operation

Iraq Enduring by carefully managing over 250 External and Stability Transition assets for the J3 United States Forces Iraq. My guidance and leadership were instrumental in navigating the complex and critical issues surrounding the Iraqi people's realization of sovereignty and the Iraqi Security Forces being thrust into the lead role for security operations. As the WMD/Terrorism/CBRNE NCOIC, I performed this task at a critical time in Iraq's evolving history, and success was built upon my superb efforts in which I mentored, coached, trained, and led by example what would become the best operating section in CENTCOM and Iraq, and in Mosul, then the most complex and lethal part of all Iraq.

He is also an adjunct professor, and teaches on the following subject matters; homeland security-related topics, terrorism, mass casualty, mass fatality, and weapons of mass destruction at Central Texas College on the Fort Leonard Wood Campus, Missouri.

Introduction

The First Responder is a designated level of emergency care provider which is an integral part of the Emergency Responds Services System.

The term "first responder" has been applied to the first individual who arrives at the scene regardless of the individual's type of credential. It is the goal of the First Responder to have the core knowledge, skills, and attitudes to function in the capacity of a first responder. The First Responder uses a limited amount of equipment to perform initial assessment and intervention and is trained to assist other providers.

It is recognized that there may be additional specific education that will be required of First Responders who operate in the field. It is also recognized that practice might differ from locality to locality, and that each training program or system should identify and provide additional training requirements to incorporate additional skills into the scope of practice of the First Responder.

Emergency responders have varying levels of familiarity with training, experience, and an agency's standard operating procedures during the response and recovery phases of an emergency situation.

The situational assessment must be made prior to any response action, and this assessment must look at what is available in the community, how we as responders react to the "All Hazards Approach", what procedures are required, and whether we have the equipment, training, and funding to perform our duties.

When a local community's resources are unable to respond or are overwhelmed, the event is then categorized as a "disaster." More victims than responder's Federal resources (which may include FEMA and DSCA) are activated. There are two basic types of events: events ***with*** notice and events ***without*** notice.

- Events with notice are disasters or emergencies where there is ***time*** to prepare and/or evacuate. Some examples are hurricanes or wildfires.

- Events without notice are disasters or emergencies where there is ***no time*** to prepare and/or evacuate. Some examples are terrorist attacks, chemical or hazardous material spills, airline accidents, or earthquakes.

These events break down into many aspects, and may fall into one or more of these categories:

Specific emergency scenes

- Specific emergency scenes—disasters disrupt hundreds of thousands of lives every year. Each disaster has lasting effects, both to people and property. Disasters ***affect all populations***. Even the emergency responders are impacted by a disaster.

Natural disasters

- Blizzards
- Earthquake
- Flood
- Hurricane
- Ice storms
- Land/mud slides
- Lightning strikes
- Tornado
- Wildfire
- Wind shears

Transportation (mass transit) emergency incidents

- Aviation
- Buses
- Seaport
- Train

Industrial/household incidents

- Electrical
- Fire
- Structural

Terrorist incidents

- Biological
- Bomb/explosion

- Chemical
- Cyber
- Ecological
- Nuclear

Hazardous materials incidents

- Biohazard
- Chemical
- Nuclear
- Human and animal disease pandemics (outbreaks)
- Search and rescue missions
- Criminal acts and crime scene investigations
- Schools violence and other emergencies

A **first responder** is a person who has completed a course and received certification in providing pre-hospital care for medical emergencies. They have more skill than someone who is trained in basic first aid, but they are not a substitute for more advanced medical care rendered by emergency medical technicians (EMTs) or paramedics. First responder courses cover the human body, lifting and moving patients, patient assessment, medical and trauma emergencies, cardiopulmonary resuscitation (CPR), automated external defibrillator usage, spinal and bone fracture immobilization, and EMS operations. The term "certified first responder" is not to be confused with "first responder," which is a generic term referring to the first medically trained responder to arrive on scene (police, fire, EMS). Most police officers and all professional firefighters in the United States and Canada, and many other countries, are certified first responders. This is the required level of training. Some police officers and firefighters obtain more training to become emergency medical technicians or paramedics. The title of first responder now covers many other professional fields to include the United States of America Department of Defense.

Limitations on certified first responders

While certified first responders are covered under Good Samaritan laws in jurisdictions where they are enacted, in some cases they have a **duty to act**. Certified first responders who are providing medical coverage to events (such as Red Cross and patient care divisions at community events), as well as those who are employed by volunteer fire departments, campus response teams, and others who are required to perform emergency medical response as part of their duties all have a duty to act.

While certified first responders in general are not required to render aid to injured/ill persons, those who work in the aforementioned areas can be accused of, and prosecuted for, negligence if they fail to respond when notified of a medical emergency, if their care does not meet the standard to which they were trained, or their care exceeds their scope of practice and causes harm to the patient.

As with all medically trained and certified persons, certified first responders are immune to successful prosecution if assistance was given in good faith up to, and not beyond, the limits of certification and training.

Scope of practice

Emergency responders are tested during a training exercise.

First responders in the United States can either provide emergency care first on the scene (police/fire department/search and rescue/park rangers) or support emergency medical technicians and paramedics. They can perform assessments, take vital signs, provide treatment for trauma and medical emergencies, perform CPR, use an automated external defibrillator, immobilize bone fractures and spinal injuries, administer oxygen, and maintain an open airway through the use of suctioning and airway adjuncts. They are permitted to assist in the administration of epinephrine auto-injectors, inhalers, and oral glucose. They are also trained in packaging, moving, and transporting patients.

First responder skills and limitations

First responder training differs per state or country. Lifesaving skills in the first responder course include recognizing unsafe scenarios and hazardous materials emergencies, protection from blood-borne pathogens, controlling bleeding, applying splints, conducting a primary life-saving patient assessment, in-line spinal stabilization and transport, CPR, and calling for more advanced medical help. Some areas give more training in other life-saving techniques and equipment.

Emergency medical oxygen is a common supplementary skill that may be added in accordance with the 1995 *DOT First Responder: National Standard Curriculum* guidelines, or under the authority of EMS agencies or training providers such as the American Red Cross. Other supplementary skills at this level can include the taking of vital signs, such as manual blood pressures, advanced splinting, and the use of the automated external defibrillator (AED), suction, and airway adjuncts.

First responders can serve as secondary providers with some volunteer EMS services. A certified first responder can be seen either as an advanced first aid provider, or as a limited provider of emergency medical care when more advanced providers are not yet on scene or available.

Rescue

The National Fire Protection Association standards 1006 and 1670 state that all "rescuers" must have medical training to perform any technical rescue operation, including cutting the vehicle itself during an extrication. Therefore, whether it is an EMS or fire department that runs the rescue, the actual rescuers who cut the vehicle and run the extrication scene or perform any rescue (e.g., rope rescues) are medical first responders, emergency medical technicians, or paramedics, as most every rescue has a patient involved.

Traditional first responders

The first responder training is considered a bare minimum for emergency service workers who may be sent out in response to a call for help. Professional firefighters require valid CFR-D (certified first responder-defibrillation) certification. The first responder level of emergency medical training is also often required for police officers. Many responders have location-specific training, such as water or mountain rescue, and must take advanced courses to be certified (e.g., ski patrol/lifeguard).

Non-traditional first responders

Many people who do not fall into the earlier-mentioned categories seek out or receive certified first responder training through their employment because they are likely to be first on the scene of a medical emergency, or because they work far from medical help.

Terrorism and the security and policies of the United States

The foundation of the United States' counterterrorism policy, according to the U.S. State Department Coordinator for Counterterrorism, are embodied in four principles following Presidential Decision Directive (PDD) 39, which ordered the Attorney General, the Director of the FBI, the Director of Central Intelligence (DCI), and the secretaries of State, Defense, Transportation, and Treasury, to enact measures to reduce vulnerabilities to terrorism. Also critical in this regard is the General Accounting Office (GAO), whose responsibilities include national preparedness, and the General Services Administration (GSA), which, as overseer of government building projects, has been increasingly tasked with providing structural protections against attacks.

The Patriot Act

The leading statement of deterrence policy since September 2001 is the Patriot Act, which President Bush signed into law on October 26, just six weeks after the attacks.

The law contained changes to some 15 different statutes, and its provisions collectively gave the Justice Department and its agencies a number of new powers in intelligence-gathering and criminal procedure against drug trafficking, immigration violations, organized criminal activity, money laundering, and terrorism and terrorism-related acts themselves.

Among its specific provisions, the Patriot Act gave increased authority to intercept communications related to an expanded list of terrorism-related crimes; allowed investigators to aggressively pursue terrorists on the internet; provided new subpoena powers to obtain financial information; reduced bureaucracy by allowing investigators to use a single court order for tracing a communication nationwide; and encouraged sharing of information between local law enforcement and the intelligence community.

The Patriot Act also provided for the creation of a "terrorist exclusion list" (TEL). Members of organizations listed on the TEL may be prevented from entering the country, and in certain circumstances may be deported. Before the Secretary of State places an organization on the TEL, he or she must find that its members commit or incite terrorist activity, gather information on potential targets for terrorist activity, or provide material support to further terrorist activity.

Assignments for specific agencies

In its provisions for responding to terrorism, PDD 39 designated the State Department as the lead agency for attacks on civilians outside of the United States. It also established the State Department Foreign Emergency Support Team (FEST) and the FBI's Domestic Emergency Support Team (DEST).

The Directive gave the Federal Aviation Administration (FAA) authority to deal with "air piracy," and assigned authority over hijackings to the Department of Justice, working in concert with the departments of State, Defense, and Transportation. That particular part of PDD 39 has been superseded by the Aviation and Transportation Security Act (ATSA). Signed into law by President Bush on November 19, 2001, the ATSA created the Transportation Security Administration (TSA), now part of the Department of Homeland Security (DHS).

In its consequence management provisions, PDD 39 gave the Federal Emergency Management Agency (FEMA) the responsibility of developing an overall federal response plan, and ensuring that states developed their own plans. This provision of PDD 39 is just one of many statements of policy on the coordination of consequence management responsibilities, which involve an array of departments, agencies, and offices, most notably FEMA, the Environmental Protection Agency (EPA), and the Coast Guard.

Nearly two decades earlier, Congress passed the Emergency Planning and Community Right-to-Know Act (EPCRA) in 1986, which established guidelines for the assistance of local communities by federal agencies in the event of a toxic chemical spill or related incident. EPCRA also provides a framework for action both by citizens and state governments. Since the time of its passage, EPCRA and similar provisions have been increasingly understood to deal also with terrorist incidents, which may involve unleashing of lethal substances.

Similarly, in 1985, a FEMA committee had drawn up the Federal Radiological Emergency Response Plan (FRERP), a blueprint for the response of the U.S. federal government to a radiological emergency—that is, a crisis involving the release of nuclear radiation. The FRERP is an agreement among 17 federal agencies, key among which are FEMA, the Nuclear Regulatory Commission (NRC), the Departments of Energy and Defense, and the EPA.

Also important is the Coast Guard, which, in addition to protecting ports and shorelines, operates the National Response Center. The latter is the sole national point of contact for reports of oil spills, as well information regarding discharges of chemical, radiological, and biological into the environment.

Agencies tasked for counterterrorism

Myriad government intelligence, security, and law enforcement agencies have a counterterrorism function. Most obvious among these are various components of the U.S. military, most notably Delta Force and Seal Team Six. These special teams, along with the larger Special Operations Command, are the "muscle" of U.S. counterterrorism. Highly trained, and well-equipped with state-of-the-art weaponry, airborne insertion equipment, and other forms of technology, elite counterterrorist teams are capable of rescuing hostages and eliminating terrorists in situations for which regular military forces would be inappropriate.

Equally vital is the work of the Coordinator for Counterterrorism. In accordance with the fourth major principle of U.S. counterterrorism policy, the Coordinator is charged by the Secretary of State with coordinating efforts to improve cooperation between the U.S. government and its foreign counterparts in battling terrorism. An ambassador, the Coordinator is the primary functionary of the federal government for developing and implementing America's counterterrorism policy.

DCI Counterterrorist Center

An entirely different wing of government is the DCI Counterterrorist Center (CTC). Though part of the CIA, the CTC is under more direct control by the DCI than are most CIA activities, a sign of the significance attached to counterterrorism. During the mid-1980s, a panel led by then-Vice President George Bush studied U.S. efforts against terrorism and concluded that, while U.S. agencies collected information on foreign terrorism, they did not aggressively operate to disrupt terrorist activities. On these recommendations from Bush, himself a former DCI, William Casey established the CTC.

The mission of the CTC is to assist the DCI in coordinating the counterterrorism efforts of the intelligence community by implementing a comprehensive counterterrorist operations program, and by exploiting all sources of intelligence to produce in-depth analyses of terror groups and their state supporters. CTC collects information on these groups, and when it has credible information of a threat, issues warnings. Alongside it is the Interagency Intelligence Committee on Terrorism, an Intelligence Community board that assists the DCI in coordinating intelligence-gathering efforts against terrorists. In the 1990s, the CTC began working closely with the FBI, and in 1996 they exchanged senior-level officers to manage the counterterrorist offices of both agencies.

The FBI

Prior to September 2001, the mission of the FBI had been strictly that of a law enforcement agency, but in the wake of September 11, Attorney General John Ashcroft and FBI Director Robert S. Mueller III refocused the bureau's efforts toward counterterrorism. In December 2001, Mueller announced plans to reorganize headquarters by creating new counterterrorism, cybercrimes, and counterintelligence divisions, by modernizing information systems, and emphasizing relationships with local first responders.

By the spring of 2002, criticism of Mueller's plans was on the rise, with detractors maintaining that the measures were not thorough enough. To this end, Mueller announced a number of new reforms. These included the hiring of 400 more analysts, including 25 from the CIA; the re-tasking of 480 special agents from white-collar and violent crimes to counterterrorism; the creation of an intelligence office; development of terrorism expert support teams to work with the bureau's 56 field offices; recruitment of Arabic speakers and others fluent in Middle Eastern and South Asian languages; creation of a joint terrorism task force to coordinate with the CIA and other federal agencies; and the improvement of financial analysis and other forms of strategic analysis directed toward terrorist groups.

In January 2003, President Bush announced plans to create a new counterterrorism intelligence center that would bring together intelligence collected domestically with that gathered overseas. This idea had been in development for some time, but one major issue of dispute was the question of which agency, the FBI or CIA, should manage the new center. One proposal put forward at the time involved the expansion of the DCI Counterterrorist Center, the oldest office of its kind. In February, Bush unveiled the organizational blueprint for the new unit, which would bring together FBI and CIA efforts under the aegis of a Terrorist Threat Integration Center, headed by the CIA.

The FBI divides the terrorist threat facing the United States into two broad categories—domestic and international.

Domestic terrorism is the unlawful use, or threatened use, of violence by a group or individual based and operating entirely within the United States (or its territories) without foreign direction, committed against persons or property to intimidate or coerce a government, the civilian population, or any segment thereof, in furtherance of political or social objectives.

International terrorism involves violent acts or acts dangerous to human life that are a violation of the criminal laws of the United States or any state, or that would be a criminal violation if committed within the jurisdiction of the United States or any state. Acts are intended to intimidate or coerce a civilian population, influence the policy of a government, or affect the conduct of a government. These acts transcend national boundaries in terms of the means by which they are accomplished, the persons they appear intended to intimidate, or the locale in which perpetrators operate. As events during the past several years demonstrate, both domestic and international terrorist organizations represent threats to Americans within the borders of the United States.

The great struggles of the twentieth century between liberty and totalitarianism ended with a decisive victory for the forces of freedom—and a single sustainable model for national success: freedom, democracy, and free enterprise. In the twenty-first century, only nations that share a commitment to protecting basic human rights and guaranteeing political and economic freedom will be able to unleash the potential of their people and assure their future prosperity. People everywhere want to be able to speak freely; choose who will govern them; worship as they please; educate their children—male and female; own property; and enjoy the benefits of their labor. These values of freedom are right and true for every person, in every society—and the duty of protecting these values against their enemies is the common calling of freedom-loving people across the globe and across the ages.

Today, the United States enjoys a position of unparalleled military strength and great economic and political influence. In keeping with our heritage and principles, we do not use our strength to press for unilateral advantage. We seek instead to create a balance of power that favors human freedom: conditions in which all nations and all societies can choose for themselves the rewards and challenges of political and economic liberty. In a world that is safe, people will be able to make their own lives better. We will defend the peace by fighting terrorists and tyrants. We will preserve the peace by building good relations among the great powers. We will extend the peace by encouraging free and open societies on every continent.

Defending our nation against its enemies is the first and fundamental commitment of the federal government. Today, that task has changed dramatically. Enemies in the past needed great armies and great industrial capabilities to endanger America. Now, shadowy networks of individuals can bring great chaos and suffering to our shores for less than it costs to purchase a single tank. Terrorists are organized to penetrate open societies and to turn the power of modern technologies against us.

To defeat this threat we must make use of every tool in our arsenal—military power, better homeland defenses, law enforcement, intelligence, and vigorous efforts to cut off terrorist financing. The war against terrorists of global reach is a global enterprise of uncertain duration. America will help nations that need our assistance in combating terror. And America will hold to account nations that are compromised by terror, including those who harbor terrorists—because the allies of terror are the enemies of civilization. The United States and countries cooperating with us must not allow the terrorists to develop new home bases. Together, we will seek to deny them sanctuary at every turn.

The gravest danger our nation faces lies at the crossroads of radicalism and technology. Our enemies have openly declared that they are seeking weapons of mass destruction, and evidence indicates that they are doing so with determination. The United States will not allow these efforts to succeed. We will build defenses against ballistic missiles and other means of delivery. We will cooperate with other nations to deny, contain, and curtail our enemies' efforts to acquire dangerous technologies. And, as a matter of common sense and self-defense, America will act against such emerging threats before they are fully formed. We cannot defend America and our friends by hoping for the best. So we must be prepared to defeat our enemies' plans, using the best intelligence and proceeding with deliberation. History will judge harshly those who saw this coming danger but failed to act. In the new world we have entered, the only path to peace and security is the path of action.

As we defend the peace, we will also take advantage of an historic opportunity to preserve the peace. Today, the international community has the best chance since the rise of the nation-state in the seventeenth century to build a world where great powers compete in peace instead of continually prepare for war. Today, the world's great powers find ourselves on the same side—united by common dangers of terrorist violence and chaos. The United States will build on these common interests to promote global security. We are also increasingly united by common values. Russia is in the midst of a hopeful transition, reaching for its democratic future and a partner in the war on terror. Chinese leaders are discovering that economic freedom is the only source of national wealth. In time, they will find that social and political freedom is the only source of national greatness. America will encourage the advancement of democracy and economic openness in both nations, because these are the best foundations for domestic stability and international order. We will strongly resist aggression from other great powers—even as we welcome their peaceful pursuit of prosperity, trade, and cultural advancement.

Finally, the United States will use this moment of opportunity to extend the benefits of freedom across the globe. We will actively work to bring the hope of democracy, development, free markets, and free trade to every corner of the world. The events of September 11, 2001, taught us that weak states, like Afghanistan, can pose as great a danger to our national interests as strong states. Poverty does not make poor people into terrorists and murderers. Yet poverty, weak institutions, and corruption can make weak states vulnerable to terrorist networks and drug cartels within their borders.

> The United States will stand beside any nation determined to build a better future by seeking the rewards of liberty for its people. Free trade and free markets have proven their ability to lift whole societies out of poverty—so the United States will work with individual nations, entire regions, and the entire global trading community to build a world that trades in freedom and therefore grows in prosperity. The United States will deliver greater development assistance through the New Millennium Challenge Account to nations that govern justly, invest in their people, and encourage economic freedom. We will also continue to lead the world in efforts to reduce the terrible toll of HIV/AIDS and other infectious diseases.
>
> In building a balance of power that favors freedom, the United States is guided by the conviction that all nations have important responsibilities. Nations that enjoy freedom must actively fight terror. Nations that depend on international stability must help prevent the spread of weapons of mass destruction. Nations

that seek international aid must govern themselves wisely, so that aid is well spent. For freedom to thrive, accountability must be expected and required.

We are also guided by the conviction that no nation can build a safer, better world alone. Alliances and multilateral institutions can multiply the strength of freedom-loving nations. The United States is committed to lasting institutions like the United Nations, the World Trade Organization, the Organization of American States, and NATO as well as other long-standing alliances. Coalitions of the willing can augment these permanent institutions. In all cases, international obligations are to be taken seriously. They are not to be undertaken symbolically to rally support for an ideal without furthering its attainment.

Freedom is the non-negotiable demand of human dignity; the birthright of every person—in every civilization.

Throughout history, freedom has been threatened by war and terror; it has been challenged by the clashing wills of powerful states and the evil designs of tyrants; and it has been tested by widespread poverty and disease. Today, humanity holds in its hands the opportunity to further freedom's triumph over all these foes. The United States welcomes our responsibility to lead in this great mission.

(George W. Bush 2007)

Conclusion

The objective of the this handbook is to make available the information to enhance and support the response capabilities of firefighters, emergency medical technicians, hazardous materials response teams, law enforcement officers, bomb squads, medical doctors and nurses, emergency managers, schools and other emergency responders at the local, state and federal levels of government, members of the public and private sector, and the citizens of this country. This mission is achieved through education, training, planning, technology, research, partnerships, community outreach, and other means.

This handbook is dedicated and committed to disaster prevention, preparedness, readiness, response, mitigation, and recovery efforts. The successful management of disasters hinges mainly on a strong and ongoing partnership among first responders, citizens and community.

chapter one

NIMS and the incident command system

Introduction

The *National Incident Management System* and the *Incident Command System* comprise an all-hazard system intended to play a role in on-scene response efforts across the United States.

The way this nation prepares for and responds to domestic incidents has changed. Best practices that have been developed over the years form part of this comprehensive national approach to incident management known as the National Incident Management System (NIMS).

Developed by the Department of Homeland Security (DHS), the NIMS will enable responders at all jurisdictional levels and across all disciplines to work together more effectively and efficiently.

One of the most important "best practices" that has been incorporated into the NIMS is the Incident Command System (ICS), a standard, on-scene, all-hazards incident-management system already in use by firefighters, hazardous materials teams, rescuers, and emergency medical teams. The ICS has been established by the NIMS as the standardized incident organizational structure for the management of all incidents.

In Homeland Security Presidential Directive-5 (HSPD-5), President Bush called on the Secretary of Homeland Security to develop a national incident management system to provide a consistent nationwide approach for federal, state, tribal, and local governments to work together to prepare for, prevent, respond to, and recover from domestic incidents, regardless of cause, size or complexity.

In collaboration with state and local government officials and representatives from a wide range of public-safety organizations, Homeland Security established the NIMS. It incorporates many existing best practices into a comprehensive national approach to domestic incident management, applicable at all jurisdictional levels and across all functional disciplines.

The NIMS represents a core set of doctrines, principles, terminology, and organizational processes to enable effective, efficient, and collaborative incident management at all levels. To provide the framework for interoperability and compatibility, the NIMS is based on a balance between flexibility and standardization. The recommendations of the National Commission on Terrorist Attacks upon the United States (the "9/11 Commission") further highlighted the importance of the ICS. The Commission's report recommended national adoption of the ICS to enhance command, control, and communications capabilities.

The history of the incident command system

The concept of an ICS was developed more than 30 years ago, in the aftermath of devastating wildfires in California. During 13 days in 1970, 16 lives were lost, 700 structures were destroyed, and over half a million acres burned. The overall cost and loss associated with these fires totaled $18 million per day. Although all of the responding agencies cooperated to the best of their ability, numerous problems with communication and coordination hampered their effectiveness. As a result, the Congress mandated that the U.S. Forest Service design a system that would "make a quantum jump in the capabilities of Southern California wild land fire protection agencies to effectively coordinate interagency action and to allocate suppression resources in dynamic, multiple-fire situations."

The California Department of Forestry and Fire Protection; the Governor's Office of Emergency Services; the Los Angeles, Ventura, and Santa Barbara County Fire Departments; and the Los Angeles City Fire Department joined with the U.S. Forest Service to develop the system. This system became known as FIRESCOPE (Firefighting Resources of California Organized for Potential Emergencies).

In 1973, the first FIRESCOPE Technical Team was established to guide the research and development design. Two major components came out of this work, the ICS and the Multi-Agency Coordination System (MACS). The FIRESCOPE ICS is primarily a command-and-control system delineating job responsibilities and organizational structure for the purpose of managing day-to-day operations for all types of emergency incidents.

By the mid-seventies, the FIRESCOPE agencies had formally agreed upon on ICS common terminology and procedures and conducted limited field-testing of ICS. By 1980, parts of ICS had been used successfully on several major wild land and urban fire incidents. It was formally adopted by the Los Angeles Fire Department, the California Department of Forestry and Fire Protection (CDF), and the Governor's Office of Emergency Services (OES), and endorsed by the State Board of Fire Services.

Also during the 1970s, the National Wildfire Coordinating Group (NWCG) was chartered to coordinate fire management programs of the various participating federal and state agencies. By 1980, FIRESCOPE ICS training was under development. Recognizing that in addition to the local users for which it was designed, the FIRESCOPE training could satisfy the needs of other state and federal agencies, the NWCG conducted an analysis of FIRESCOPE ICS for possible national application.

By 1981, ICS was widely used throughout Southern California by the major fire agencies. In addition, the use of ICS in response to nonfire incidents was increasing. Although FIRESCOPE ICS was originally developed to assist in the response to wild land fires, it was quickly recognized as a system that could help public-safety responders provide effective and coordinated incident management for a wide range of situations, including floods, hazardous-materials accidents, earthquakes, and aircraft crashes.

It was flexible enough to manage catastrophic incidents involving thousands of emergency response and management personnel. By introducing relatively minor terminology, organizational, and procedural modifications to FIRESCOPE ICS, the NIMS ICS became adaptable to an all-hazards environment.

While tactically, each type of incident may be handled somewhat differently, the overall incident-management approach still utilizes the major functions of the ICS. In 1982, after the FIRESCOPE board of directors and the NWCG recommended its national application, all FIRESCOPE ICS documentation was revised and adopted as the National Interagency Incident Management System (NIIMS). In the years since FIRESCOPE and the NIIMS were blended, the FIRESCOPE agencies and the NWCG have worked together to update and maintain the Incident Command System Operational System Description (ICS 120-1). This document would later serve as the basis for the NIMS ICS.

National Incident Management System

The NIMS provides a consistent, flexible, and adjustable national framework within which government and private entities at all levels can work together to manage domestic incidents, regardless of their cause, size, location, or complexity. This flexibility applies across all phases of incident management: prevention, preparedness, response, recovery, and mitigation.

The NIMS provides a set of standardized organizational structures including the ICS, MACS, and public information systems as well as requirements for processes, procedures, and systems to improve interoperability among jurisdictions and disciplines in various areas.

Homeland Security recognizes that the overwhelming majority of emergency incidents are handled on a daily basis by a single jurisdiction at the local level. However, the challenges we face as a nation are far greater than the capabilities of any one community or state, but not greater than the sum of us all working together.

There will be instances in which successful domestic incident-management operations depend on the involvement of emergency responders from multiple jurisdictions, as well as personnel and equipment from other states and the federal government. These instances require effective and efficient coordination across a broad spectrum of organizations and activities.

The success of the operations will depend on the ability to mobilize and effectively utilize multiple outside resources. These resources must come together in an organizational framework that is understood by everyone and must utilize a common plan, as specified through a process of incident action planning. This will only be possible if we unite, plan, exercise, and respond using a common National Incident Management System.

What is NIMS ICS?

With the exception of the way the intelligence function is handled, the principles and concepts of NIMS/ICS are the same as for FIRESCOPE and NIMS/ICS.

ICS management

The characteristic ICS approach is based on proven management tools that contribute to the strength and efficiency of the overall system. The following ICS management characteristics are taught by the DHS in its ICS training programs:

- Common terminology
- Modular organization
- Management by objectives
- Reliance on an incident action plan
- Manageable span of control
- Pre-designated incident mobilization-center locations and facilities
- Comprehensive resource management
- Integrated communications
- Establishment and transfer of command
- Chain of command and unity of command
- Unified command
- Accountability of resources and personnel
- Deployment
- Information and intelligence management.

ICS Command comprises the Incident Commander (IC) and Command Staff. Command staff positions are established to assign responsibility for key activities not specifically identified in the General Staff functional elements. These positions may include the Public Information Officer (PIO), Safety Officer (SO), and the Liaison Officer (LNO), in additional to various others, as required and assigned by the IC.

Unified Command (UC) is an important element in multijurisdictional or multiagency domestic incident management. It provides guidelines to enable agencies with different legal, geographic, and functional responsibilities to coordinate, plan, and interact effectively.

The Unified Command team overcomes much of the inefficiency and duplication of effort that can occur when agencies from different functional and geographic jurisdictions, or agencies at different levels of government, operate without a common system or organizational framework. The primary difference between the single command structure and the UC structure is that in a single command structure, the IC is solely responsible for establishing incident-management objectives and strategies. In a UC structure, the individuals designated by their jurisdictional authorities jointly determine objectives, plans, and priorities and work together to execute them.

General Staff includes incident-management personnel who represent the major functional elements of the ICS, including the Operations Section Chief, Planning Section Chief, Logistics Section Chief, and Finance/Administration Section Chief. Command Staff and General Staff must continually interact and share vital information and estimates of the current and future situation and develop recommended courses of action for consideration by the IC.

The *Incident Action Plan* (IAP) includes the overall incident objectives and strategies established by the IC or UC. The Planning Section is responsible for developing and documenting the IAP. In the case of UC, the IAP must adequately address the overall incident objectives, mission, operational assignments, and policy needs of each jurisdictional agency. This planning process is accomplished with productive interaction between jurisdictions, functional agencies, and private organizations. The IAP also addresses tactical objectives and support activities for one operational period, generally 12–24 h. The IAP also contains provisions for continuous incorporation of "lessons learned" as identified by the Incident Safety Officer or incident-management personnel as activities progress.

Area Command is activated only if necessary, depending on the complexity of the incident and span-of-control considerations. An Area Command is established either to oversee the management of multiple incidents that are being handled by separate ICS organizations or to oversee the management of a very large incident that involves multiple ICS organizations. It is important to note that Area Command does not have operational responsibilities. For incidents under its authority, the Area Command:

- Sets overall agency incident-related priorities
- Allocates critical resources according to established priorities

- Ensures that incidents are managed properly
- Ensures effective communications
- Ensures that incident-management objectives are met and do not conflict with each other or with agency policies; identifies critical resource needs and reports them to emergency operations
- Center(s): ensures that short-term emergency recovery is coordinated to assist in the transition to full recovery operations; and provides for personnel accountability and a safe operating environment

Available NIMS ICS training

DHS, through its many training bodies, makes ICS training available. ICS training developed by the Federal Emergency Management Agency (FEMA) includes:

- ICS-100, Introduction to ICS
- ICS-200, Basic ICS Single Resources and Initial Action Incidents
- ICS-300, Intermediate ICS for Expanding Incidents
- ICS-400, Advanced ICS Command and General Staff, Complex Incidents
- There are multiple other NIMS/ICS courses online that are free from FEMA

To participate in FEMA's ICS training, contact the state emergency management training office. The Emergency Management Institute (EMI) and the National Fire Academy (NFA) also offer ICS Train-the-Trainer classes at their Emmetsburg, MD, facility.

A variety of other ICS training programs are available. The NIMS Integration Center is working with federal and state training providers to ensure that their ICS course offerings are consistent with the NIMS. http://www.fema.gov/incident-command-system

It is important to remember why we have the NIMS and why ICS is a critical piece of the incident-management system. Most incidents are local, but when faced with the worst-case scenario, such as September 11, 2001, all responding agencies must be able to interface and work together. The NIMS, in particular the ICS component, allows that to happen, but only if the foundation has been laid at the local level. If local jurisdictions adopt a variation of ICS that cannot grow or is not applicable to other disciplines, the responding agencies and jurisdictions cannot interface in an expanded response.

It is important that everyone understand that with the establishment of the NIMS, there is only one ICS. As agencies adopt the principles and concepts of ICS as established in the NIMS, the incident command system can expand to meet the needs of the response, regardless of the size or number of responders. The key to both NIMS and ICS is a balance between standardization and flexibility. The NIMS document is available on www.fema.gov/nims.

There are essential elements to the ICS, and these elements keep the system simple and user-friendly at any level when being implemented for any type of event or incident.

The 14 essential, standardized ICS features are listed below.

1. *Common terminology*: Using common terminology helps to define organizational functions, incident facilities, resource descriptions, and position titles.
2. *Command*:

   Establishment and transfer of command; the command function must be clearly established from the beginning of an incident. When command is transferred, the process must include a briefing that captures all essential information for continuing safe and effective operations.

   Chain of command and unity of command; chain of command refers to the orderly line of authority within the ranks of the incident-management organization. Unity of command means that every individual has a designated supervisor to whom he or she reports at the scene of the incident.

   These principles clarify reporting relationships and eliminate the confusion caused by multiple, conflicting directives. Incident managers at all levels must be able to control the actions of all personnel under their supervision.

   Unified Command; in incidents involving multiple jurisdictions, a single jurisdiction with multiagency involvement, or multiple jurisdictions with multiagency involvement, Unified Command allows agencies with different legal, geographic, and functional authorities and responsibilities to work together effectively without affecting individual agency authority, responsibility, or accountability.
3. *Planning/organizational structure*:

   Management by objectives; includes establishing overarching objectives; developing strategies based on incident objectives; developing and issuing assignments, plans, procedures, and protocols; establishing specific, measurable objectives for various incident-management functional activities and directing efforts to attain them, in support of defined strategies; and documenting results to measure performance and facilitate corrective action.

   Modular organization; the Incident Command organizational structure develops in a modular fashion that is based on the size and complexity of the incident, as well as the specifics of the hazard environment created by the incident.

Incident action planning; incident action plans (IAPs) provide a coherent means of communicating the overall incident objectives in the context of both operational and support activities.

Manageable span-of-control; span of control is key to effective and efficient incident management. Within ICS, the span of control of any individual with incident management supervisory responsibility should range from three to seven subordinates.

4. *Facilities and Resources*:

   Incident locations and facilities; various types of operational support facilities are established in the vicinity of an incident to accomplish a variety of purposes. Typical designated facilities include incident command posts, bases, camps, staging areas, mass casualty triage areas, and others as required.

   Comprehensive resource management; maintaining an accurate and up-to-date picture of resource utilization is a critical component of incident management. Resources are defined as personnel, teams, equipment, supplies, and facilities available or potentially available for assignment or allocation in support of incident management and emergency response activities.

5. *Communications/information management*:

   Integrated communications; incident communications are facilitated through the development and use of a common communications plan and interoperable communications processes and architectures.

   Information and intelligence management; the incident-management organization must establish a process for gathering, analyzing, sharing, and managing incident-related information and intelligence.

6. *Professionalism*
7. *Accountability*: effective accountability at all jurisdictional levels and within individual functional areas during incident operations is essential. To that end, the principles in the points which follow must be adhered to.
8. *Check-in*: all responders, regardless of agency affiliation, must report in to receive an assignment in accordance with the procedures established by the Incident Commander.
9. *Incident action plan*: response operations must be directed and coordinated as outlined in the IAP.
10. *Unity of command*: each individual involved in incident operations will be assigned to only one supervisor.
11. *Personal responsibility*: all responders are expected to use good judgment and be accountable for their actions.

12. *Span of control*: supervisors must be able to adequately supervise and control their subordinates, as well as communicate with and manage all resources under their supervision.
13. *Resource tracking*: supervisors must record and report resource status changes as they occur.
14. *Dispatch/deployment*: personnel and equipment should respond only when requested or when dispatched by an appropriate authority.

This list is a basic descriptive break down of the ICS forms used during an event or incident.

The ICS uses a series of standard forms and supporting documents that convey directions for the accomplishment of the objectives and distributing information. Listed below are the standard ICS form titles and descriptions of each form.

Standard form title	Description
Incident action plan cover page ICS 200	Indicates the incident name, plan operational period, date prepared, approvals, and attachments (resources, organization, communications plan, medical plan, and other appropriate information).
Incident briefing ICS 201	Provides the Incident Command/Unified Command and General Staffs with basic information regarding the incident situation and the resources allocated to the incident. This form also serves as a permanent record of the initial response to the incident.
Incident objectives ICS 202	Describes the basic strategy and objectives for use during each operational period.
Organization assignment list ICS 203	Provides information on the response organization and personnel staffing.
Field assignment ICS 204	Used to inform personnel of assignments. After Incident Command/Unified Command approves the objectives, staff members receive the assignment information contained in this form.
Incident communications plan ICS 205	Provides, in one location, information on the assignments for all communications equipment for each operational period. The plan is a summary of information. Information from the Incident Communications Plan on frequency assignments can be placed on the appropriate assignment form (ICS Form 204).
Medical plan ICS 206	Provides information on incident medical-aid stations, transportation services, hospitals, and medical emergency procedures.

(*Continued*)

(Continued)

Standard form title	Description
Incident status summary ICS 209	Summarizes incident information for staff members and external parties, and provides information to the Public Information Officer for preparation of media releases.
Check-in/out list ICS 211	Used to check-in personnel and equipment arriving at or departing from the incident. Check-in/out consists of reporting specific information that is recorded on the form.
General message ICS 213	Used by: Incident dispatchers to record incoming messages that cannot be orally transmitted to the intended recipients. Emergency Operation Center (EOC) and other incident personnel to transmit messages via radio or telephone to the addressee. Incident personnel to send any message or notification that requires hard-copy delivery to other incident personnel.

Conclusion

The National Incident Management System (NIMS) is a standardized approach to incident management developed by the Department of Homeland Security, and is intended to facilitate coordination between all responders (including all levels of government with public, private, and nongovernmental organizations). NIMS standard incident command structures are based on three key organizational systems:

- The Incident Command System (ICS)
- The Multiagency Coordination System
- Public Information Systems

NIMS will enable responders at all levels to work together more effectively and efficiently to manage domestic incidents no matter what the cause, size, or complexity, including catastrophic acts of terrorism and disasters. Federal agencies also are required to use the NIMS framework in domestic incident management and in support of state and local incident response and recovery activities.

chapter two

Security management

Introduction

Security management is the identification of an organization's assets (including information assets), followed by the development, documentation, and implementation of policies and procedures for protecting these assets.

An organization uses such security-management procedures as information classification, risk assessment, and risk analysis to identify threats, categorize assets, and rate system vulnerabilities so that they can implement effective controls.

Loss prevention focuses on what your critical assets are and how you are going to protect them. A key component to loss prevention is assessing potential threats to successful achievement of the goal. This must include potential opportunities to further the object (why take the risk unless there's an upside?); balance probability; and impact, determine, and implement measures to minimize or eliminate those threats.

Security risk management applies the principles of risk management to the management of security threats. It consists of identifying threats (or risk causes), assessing the effectiveness of existing controls to face those threats, determining the risks' consequence(s), prioritizing the risks by rating the likelihood and impact, classifying the type of risk, and selecting an appropriate risk option or risk response.

Types of security threats

External

- Strategic—competition and customer demand
- Operational—regulation, suppliers, contracts
- Financial—FX, credit
- Hazard—natural disaster, cyber, external criminal act
- Compliance—new regulatory or legal requirements are introduced, or existing ones are changed, exposing the organization to a non-compliance risk if measures are not taken to ensure compliance

Internal

- Strategic—R&D
- Operational—systems and process (human resources, payroll)
- Financial—liquidity, cash flow
- Hazard—safety and security; employees and equipment
- Compliance—actual or potential changes in the organization's systems, processes, suppliers, and so on may create exposure to legal or regulatory non-compliance.

Risk

Risk avoidance

Eliminating the existence of criminal opportunity or avoiding the creation of such an opportunity is always the best solution, when greater risks or other considerations are not created as a result of these actions.

As an example, removing all the cash from a retail outlet would eliminate the opportunity for stealing the cash—but it would also eliminate the ability to conduct business.

Risk reduction

When avoiding or eliminating criminal opportunity conflicts with the ability to conduct business, the next option is the reduction of the opportunity and potential loss to the lowest level consistent with the function of the business. In the example above, the application of risk reduction might result in the business keeping only enough cash on hand for one day's operation.

Risk spreading

Assets that remain at risk after reduction and avoidance are the subjects of risk spreading. This concept limits loss or potential losses by exposing the perpetrator to the probability of detection and apprehension prior to the consummation of the crime through the application of perimeter lighting, barred windows, and intrusion detection systems. The idea here is to reduce the time available to steal assets and escape without apprehension.

Risk transfer

Transferring risk to other alternatives when the risk has not been reduced to acceptable levels may be accomplished by insuring the asset or raising prices to cover loss in the event of a criminal act. Generally speaking, when the first three steps have been properly applied, the costs of transferring risks are much lower.

Risk acceptance

All remaining risks must simply be assumed by the business as a risk of doing business. Included in these accepted losses are deductibles made as part of the insurance coverage.

Security policy implementations

Intrusion detection

Alarm devices or systems of alarm devices give an audible, visual or other form of alarm signal about a problem or condition. Alarm devices are often outfitted with a siren.

Alarm devices include:

- Burglar alarms, designed to warn of burglaries. This is often a silent alarm: the police or guards are warned without indication to the burglar, which increases the chances of catching him or her.
- Alarm clocks, which can produce an alarm at a given time.
- Distributed control systems (DCS), found in nuclear power plants, refineries and chemical facilities also generate alarms to direct the operator's attention to an important event that he or she needs to address.
- Alarms in an operation and maintenance (O&M) monitoring system, which inform of the malfunction of (a particular part of) the system under monitoring.
- First-out alarms

Safety alarms go off if a dangerous condition occurs. Common public safety alarms include:

- Civil defense sirens, also known as *tornado sirens* or *air raid sirens*
- Fire alarm systems
- Fire alarm notification appliance
- "Multiple-alarm fire," a locally-specific measure of the severity of a fire and the fire department reaction required.
- Smoke detector
- Car alarms
- Autodialed alarm, also known as *community alarm*
- Personal alarm
- Tocsins, an historical method of raising an alarm

Alarms are capable of causing a fight-or-flight response in humans: a person in this mindset will panic and either flee the perceived danger or attempt to eliminate it, often ignoring rational thought in either case. We can characterize a person in such a state as "alarmed."

With any kind of alarm, the need exists to balance between, on the one hand, the danger of false alarms (called "false positives")—the signal going off in the absence of a problem—and on the other hand, failing to signal an actual problem (called a "false negative"). False alarms can be an expensive waste of resources and can even be dangerous. For example, false alarms of a fire can waste manpower, making firefighters unavailable for a real fire, and risk injury to them and others as the fire engines race to the alleged fire's location.

In addition, false alarms may acclimatize people to ignore alarm signals, and thus possibly to ignore an actual emergency—Aesop's fable of *The Boy Who Cried Wolf* exemplifies this problem.

Access control

- Locks, simple or sophisticated, such as biometric authentication and keycard locks
- A *lock* is a mechanical or electronic fastening device that is released by a physical object (such as a key, keycard, fingerprint, RFID card, security token, etc.), by supplying secret information (such as a key code or password), or by a combination thereof.
- *Biometrics* are metrics of human characteristics and traits. Biometrics (or biometric authentication) are used in computer science as a form of identification and access control. They may also be used to identify individuals in groups that are under surveillance.
- *Biometric identifiers* are the distinctive, measurable characteristics used to label and describe individuals. Biometric identifiers are often categorized as physiological versus behavioral characteristics.
- *Physiological* characteristics are related to the shape of the body. Examples include, but are not limited to, fingerprint, palm veins, face recognition, DNA, palm print, hand geometry, iris recognition, retina, and odor/scent. *Behavioral* characteristics are related to the pattern of behavior of a person, including but not limited to typing rhythm, gait, and voice.
- A *keycard lock* is a lock operated by a *keycard*, a flat, rectangular plastic card with identical dimensions to that of a credit card or American and EU driver's license which stores a physical or digital signature which the door mechanism accepts before disengaging the lock.
- There are several popular type of keycards in use including the mechanical hole card; barcode-, magnetic stripe-, and Wiegand wire-embedded cards, smart cards (with an embedded read/write electronic microchip), and RFID proximity cards.
- Keycards are frequently used in hotels as an alternative to mechanical keys.

Physical security

- Attack dogs are trained to defend or attack a territory, property, or persons either on command, on sight, or by inferred provocation. Attack dogs have been used often throughout history and are now employed in security, police, and military roles.
- A barricade, from the French *barrique* (barrel), is any object or structure that creates a barrier or obstacle to control, block passage, or force the flow of traffic in the desired direction. Adopted as a military term, a barricade denotes any improvised field fortification, most notably on city streets during urban warfare.
- Barricades also include temporary traffic barricades designed with the goal of dissuading passage into a protected or hazardous area, or large slabs of cement whose goal is to actively prevent forcible passage by a vehicle. Stripes on barricades and panel devices slope downward in the direction traffic must travel.
- There are also pedestrian barricades—sometimes called bike rack barricades, for their resemblance to a now obsolete form of bicycle stand, or police barriers. They originated in France approximately 50 years ago and are now produced around the world. They were first produced in the United States 40 years ago by Friedrichs Mfg. for New Orleans's Mardi Gras parades.
- Anti-vehicle barriers and blast barriers are sturdy barricades that can respectively counter vehicle and bomb attacks. Recently, movable blast barriers that can be used to protect humanitarian relief workers, and villagers and their homes in unsafe areas have been designed by Nanyang Technological University.
- *Security guards* or *security officers* (armed or unarmed) with wireless communication devices (e.g., two-way radio) are paid to protect property, assets, or people. They are usually privately and formally employed civilian personnel. Security officers are generally uniformed and act to protect property by maintaining a high-visibility presence to deter illegal and inappropriate actions, observing (either directly, through patrols, or by watching alarm systems or video cameras) for signs of crime, fire, or disorder, then taking action and reporting any incidents to their client and emergency services as appropriate. Until the 1980s, the term *watchman* was more commonly applied to this function, a usage dating back to at least the Middle Ages in Europe. This term was carried over to North America, where it was interchangeable with *night-watchman* until both terms were replaced with the modern security-based titles. Security guards are sometimes regarded as fulfilling a private policing function.
- Security lighting in the field of physical security is often used as a preventive and corrective measure against intrusions or other

criminal activity on a physical piece of property. Security lighting may be provided to aid in the detection of intruders, to deter intruders, or in some cases simply to increase the feeling of safety. Lighting is integral to crime prevention through environmental design.

Law enforcement agency (LEA) jurisdiction procedures

- Law enforcement agencies (LEAs) which have their ability to apply their powers restricted in some way are said to operate within a jurisdiction.

Non-executive powers jurisdictional coverage of Europol

- LEAs will have some form of geographic restriction on their ability to apply their powers. The LEA might be able to apply its powers: within a country, for example the United States of America's Bureau of Alcohol, Tobacco, Firearms and Explosives; within a division of a country, for example the Australian state Queensland Police; or across a collection of countries, for example international organizations such as Interpol, or the European Union's Europol.
- LEAs which operate across a collection of countries tend to assist in law enforcement activities, rather than directly enforcing laws, by facilitating the sharing of information necessary for law enforcement between LEAs within those countries, for example Europol has no executive powers.
- Sometimes an LEA's jurisdiction is determined by the complexity or seriousness of the non-compliance with a law. Some countries determine the jurisdiction in these circumstances by means of policy and resource allocation between agencies: in Australia, the Australian Federal Police take on complex serious matters referred to them by an agency and the agency will undertake its own investigations of less serious or complex matters by consensus. Other countries have laws which decide the jurisdiction, for example, in the United States of America some matters are required by law to be referred to other agencies if they are of a certain level of seriousness or complexity. For example, cross-state-boundary kidnapping in the United States is escalated to the Federal Bureau of Investigation. Differentiation of jurisdiction based on the seriousness and complexity of the noncompliance, either by law or by policy and consensus, can coexist in countries.
- An LEA which has a wide range of powers but whose ability is restricted geographically, typically to an area which is only part of a country, is typically referred to as *local police* or *territorial police*. Other LEAs have a jurisdiction defined by the type of laws they enforce or assist in enforcing. For example, Interpol does not work with political, military, religious, or racial matters.

- An LEA's jurisdiction usually also includes the governing bodies they support, and the LEA itself.

Fraud management

- Fraud is a deception deliberately practiced in order to secure unfair or unlawful gain (adjectival form fraudulent; to defraud is the verb). As a legal construct, fraud is both a civil wrong (i.e., a fraud victim may sue the fraud perpetrator to avoid the fraud and/or recover monetary compensation) and a criminal wrong (i.e., a fraud perpetrator may be prosecuted and imprisoned by governmental authorities). Defrauding people or organizations of money or valuables is the usual purpose of fraud, but it sometimes instead involves obtaining benefits without actually depriving anyone of money or valuables, such as obtaining a driver's license by way of false statements made in an application for the same.
- A hoax is a distinct concept that involves deception without the intention of gain or of materially damaging or depriving the victim.

Alarm management

Alarm management describes the application of human factors (or ergonomics, as the field is referred to outside the United States), along with instrumentation engineering and systems thinking, to manage the design of an alarm system to increase its usability. Most often, the major usability problem is that too many alarms are annunciated in a plant upset, commonly referred to as alarm flood (similar to an interrupt storm), since it is so similar to a flood caused by excessive rainfall input with a basically fixed drainage/output capacity. However, there may be other problems with an alarm system, such as poorly designed alarms, improperly set alarm points, ineffective annunciation, and unclear alarm messages.

Concepts

The fundamental purpose of alarm annunciation is to alert the operator to abnormal operating situations. The ultimate objective is to prevent, or at least minimize, physical and economic loss through operator intervention in response to the condition that was alarmed. For most digital control system users, losses can result from situations that threaten environmental safety, personnel safety, equipment integrity, economy of operation, and product quality control as well as plant throughput. A key factor in operator response effectiveness is the speed and accuracy with which the operator can identify alarms that require immediate action.

By default, the assignment of alarm trip points and alarm priorities constitute basic alarm management. Each individual alarm is designed to provide an alert when that process indication deviates from normal. The main problem with basic alarm management is that these features are static. The resultant alarm annunciation does not respond to changes in the mode of operation or the operating conditions.

When a major piece of process equipment like a charge pump, compressor, or fired heater shuts down, many alarms become unnecessary. These alarms are no longer independent exceptions from normal operation. They indicate, in that situation, secondary, non-critical effects and no longer provide the operator with important information. Similarly, during startup or shutdown of a process unit, many alarms are not meaningful.

This is often the case because the static alarm conditions conflict with the required operating criteria for startup and shutdown.

In all cases of major equipment failure, startups, and shutdowns, the operator must search alarm annunciation displays and analyze which alarms are significant. This wastes valuable time when the operator needs to make important operating decisions and take swift action. If the resultant flood of alarms becomes too great for the operator to comprehend, then the basic alarm management system has failed as a system that allows the operator to respond quickly and accurately to the alarms that require immediate action. In such cases, the operator has virtually no chance to minimize, let alone prevent, a significant loss.

In short, one needs to extend the objectives of alarm management beyond the basic level. It is not sufficient to utilize multiple priority levels because priority itself is often dynamic. Likewise, alarm disabling based on unit association, or suppressing audible annunciation based on priority do not provide dynamic, selective alarm annunciation. The solution must be a system that can dynamically filter the process alarms based on the current plant operation and conditions so that only the currently significant alarms are annunciated.

The fundamental purpose of dynamic alarm annunciation is to alert the operator to relevant abnormal operating situations. They include situations that have a necessary or possible operator response to ensure:

- Personnel and Environmental Safety,
- Equipment Integrity,
- Product Quality Control.

The ultimate objectives are no different from the previous basic alarm annunciation management objectives. Dynamic alarm annunciation management focuses the operator's attention by eliminating extraneous alarms, providing better recognition of critical problems, and insuring swifter, more accurate operator response.

The need for alarm management

Alarm management is usually necessary in a process manufacturing environment that is controlled by an operator using a control system, such as a DCS or a programmable logic controller (PLC). Such a system may have hundreds of individual alarms that up until very recently have probably been designed with only limited consideration of other alarms in the system. Since humans can only do one thing at a time and can pay attention to a limited number of things at a time, there needs to be a way to ensure that alarms are presented at a rate that can be assimilated by a human operator, particularly when the plant is upset or in an unusual condition. Alarms also need to be capable of directing the operator's attention to the most important problem that he or she needs to act upon, using a priority to indicate degree of importance or rank, for instance.

Some improvement methods

The techniques for achieving rate reduction range from the extremely simple ones of reducing nuisance and low-value alarms to redesigning the alarm system in a holistic way that considers the relationships between individual alarms.

Nuisance reduction

The first step in a continuous improvement program is often to measure alarm rate, and resolve any chronic problems such as alarms that have no use (often described as those that do not require the operator to take an action). Note that the alarm rate, as measured by the alarms in a journal, is not necessarily the alarm rate seen by operators. This is because a chattering alarm will not update an unacknowledged alarm annunciator (and therefore looks like a single alarm to an operator), but will show up multiple times in a journal.

Design guide

This step involves documenting the methodology or philosophy of how to design alarms. It can include things such as what to alarm, standards for alarm annunciation and text messages, and how the operator will interact with the alarms.

Documentation and rationalization

This phase is a detailed review of all alarms to document their design purpose, and to ensure that they are selected and set properly and meet the design criteria. Ideally, though not always, this stage will result in a reduction of alarms.

Advanced methods

The above steps will often still fail to prevent an alarm flood in an operational upset, so advanced methods such as alarm suppression under certain circumstances are then necessary.

As an example, shutting down a pump will always cause a low-flow alarm on the pump outlet flow, so this alarm may be suppressed if the pump was shut down, since it adds no value for the operator, because he or she already knows it was caused by the pump being shut down. This technique can of course get very complicated and requires considerable care in design. In the above case for instance, it can be argued that the low-flow alarm does add value as it confirms to the operator that the pump has indeed stopped.

Alarm management becomes more necessary as the complexity and size of manufacturing systems increases. The need for alarm management is largely because alarms can be configured on a DCS at nearly zero incremental cost. In the past, on physical control panel systems that consisted of individual pneumatic or electronic analog instruments, each alarm required expenditure and control panel real estate, so more thought usually went into the need for an alarm. Numerous disasters such as Three Mile Island, Chernobyl, and the Deepwater Horizon have established a clear need for alarm management.

The seven steps to alarm management

Step 1: Create and adopt an alarm philosophy

A comprehensive design and guideline document is produced which defines a plant standard employing a best-practice alarm management methodology.

Step 2: Alarm performance benchmarking

Analyze the alarm system to determine its strengths and deficiencies, and effectively map out a practical solution to improve it.

Step 3: "Bad actor" alarm resolution

From experience, it is known that around half of the entire alarm load usually comes from relatively few alarms. The methods for making them work properly are documented, and can be applied with minimum effort and maximum performance improvement.

Step 4: Alarm documentation and rationalization (D&R)

A full overhaul of the alarm system to ensure that each alarm complies with the alarm philosophy and the principles of good alarm management.

Step 5: Alarm system audit and enforcement

DCS alarm systems are notoriously easy to change and generally lack proper security. Methods are needed to ensure that the alarm system does not drift from its rationalized state.

Step 6: Real-time alarm management

More advanced alarm management techniques are often needed to ensure that the alarm system properly supports, rather than hinders, the operator in all operating scenarios. These include alarm shelving, state-based alarming, and alarm flood suppression technologies.

Step 7: Control and maintain alarm system performance

Proper management of change, longer-term analysis, and KPI monitoring are needed to ensure that the gains achieved from performing the steps above do not dwindle over time. Otherwise they will; the principle of "entropy" definitely applies to an alarm system.

Information technology (IT) risk

The creation of the relatively new term *information technology risk* reflects an increasing awareness that information security is simply one facet of a multitude of risks that are relevant to IT and the real-world processes it supports. Generally speaking, risk is the product of the likelihood of an event occurring and the impact that event would have on an information technology asset, i.e., risk = likelihood impact.

Further, the impact of an event on an information asset is usually taken to be the product of vulnerability in the asset and the asset's value to its stakeholders. Thus, IT risk can be expanded to (risk = threat × vulnerability × asset value).

IT risk: the potential that a given threat will exploit vulnerabilities of an asset or group of assets and thereby cause harm to the organization. It is measured in terms of a combination of the probability of occurrence of an event and its consequence.

Committee on National Security Systems

The Committee on National Security Systems of United States of America defined *risk* in different documents:

- From CNSS Instruction No. 4009 dated 26 April 2010 the basic and more technical focused definition:
- *Risk—possibility that a particular threat will adversely impact an IS by exploiting a particular vulnerability.*

- National Security Telecommunications and Information Systems Security Instruction (NSTISSI) No. 1000 introduces a probability aspect, quite similar to NIST SP 800-30 one:
  Risk—a combination of the likelihood that a threat will occur, the likelihood that a threat occurrence will result in an adverse impact, and the severity of the resulting impact

The National Information Assurance Training and Education Center defines risk in the IT field as:

- The loss potential that exists as the result of threat-vulnerability pairs. Reducing either the threat or the vulnerability reduces the risk.
- The uncertainty of loss expressed in terms of probability of such loss.
- The probability that a hostile entity will successfully exploit a particular telecommunications or COMSEC system for intelligence purposes; its factors are threat and vulnerability.
- A combination of the likelihood that a threat shall occur, the likelihood that a threat occurrence shall result in an adverse impact, and the severity of the resulting adverse impact.
- The probability that a particular threat will exploit a particular vulnerability of the system.

Measuring IT risk

You can't effectively and consistently manage what you can't measure, and you can't measure what you haven't defined.

At this point, it is useful to introduce related terms used to properly measure IT risk.

An *information security event* is an identified occurrence of a system, service or network state indicating a possible breach of information security policy or failure of safeguards, or a previously unknown situation that may be relevant to security.

Occurrence of a particular set of circumstances

- *The event can be certain or uncertain.*
- *The event can be a single occurrence or a series of occurrences. (ISO/IEC Guide 73)*

An *information security incident* is indicated by a single or a series of unwanted information security events that have a significant probability of compromising business operations and threatening information security, or events that have an actual or potential adverse effect on the security or performance of a system.

The result of an unwanted incident is termed its *impact*.

A *consequence* is the outcome of an event. There can be more than one consequence of each event. Consequences can range from positive to negative, and they can be expressed qualitatively or quantitatively.

Risk (R) is the product of the likelihood (L) of a security incident occurring times the impact (I) that the organization will incur due to the incident, that is,

$$R = L \times I.$$

The likelihood of a security incident occurrence is a function of the likelihood that a threat appears and the likelihood that the threat can exploit the relevant system vulnerabilities.

The consequences of a security incident occurrence are a function of its likely impact on the organization due to harm to the organization's assets. *Harm* is related to the value of the asset to the organization; the same asset can have different values to different organizations.

It may be seen, therefore, that R is a function of four factors:

- A = value of the assets
- T = the likelihood of the threat
- V = the nature of the vulnerability; i.e., the likelihood of exploitation (proportional to the potential benefit for the attacker and inversely proportional to the cost of exploitation)
- I = the likely impact, or extent of the harm

It may be possible to express risk in numeric or monetary values and thereby compare the cost of countermeasures to the residual risk after applying the security control. This may not always be practicable, so as the first step of risk evaluation, risks are graded dimensionlessly using three- or five-step scales.

The Open Web Application Security Project (OWASP) proposes practical risk measurement guidelines based on estimation of (i) likelihood, (ii) technical impact, and (iii) business impact. Each of these three estimations is expressed as the mean of a range of factors, scored on a scale from zero to nine.

Estimation of likelihood

The estimation of likelihood first considers certain threat factors.

- Skill level: How technically skilled (scores in parentheses) is this group of threat agents? No technical skills (1), some technical skills (3), advanced computer user (4), network and programming skills (6), security penetration skills (9)

- Motive: How motivated is this group of threat agents to find and exploit this vulnerability? Low or no reward (1), possible reward (4), high reward (9)
- Opportunity: What resources and opportunity are required for this group of threat agents to find and exploit this vulnerability? Full access or expensive resources required (0), special access or resources required (4), some access or resources required (7), no access or resources required (9)
- Size: How large is this group of threat agents? Developers (2), system administrators (2), intranet users (4), partners (5), authenticated users (6), anonymous internet users (9)

The next set of factors considered is related to the vulnerability involved, with the goal being estimate the likelihood of the particular vulnerability being discovered and exploited.

- Ease of discovery: How easy is it for this group of threat agents to discover this vulnerability? Practically impossible (1), difficult (3), easy (7), automated tools available (9)
- Ease of exploit: How easy is it for this group of threat agents to actually exploit this vulnerability? Theoretical (1), difficult (3), easy (5), automated tools available (9)
- Awareness: How well known is this vulnerability to this group of threat agents? Unknown (1), hidden (4), obvious (6), public knowledge (9)
- Intrusion detection: How likely is an exploit to be detected? Active detection in application (1), logged and reviewed (3), logged without review (8), not logged (9)

Estimation of technical impact

Technical impact can be broken down into factors aligned with the traditional security areas of concern: confidentiality, integrity, availability, and accountability. The goal is to estimate the magnitude of the impact on the system if the vulnerability were to be exploited.

- Loss of confidentiality: How much data could be disclosed and how sensitive is it? Minimal non-sensitive data disclosed (2), minimal critical data disclosed (6), extensive non-sensitive data disclosed (6), extensive critical data disclosed (7), all data disclosed (9)
- Loss of integrity: How much data could be corrupted and how damaged is it? Minimal slightly corrupt data (1), minimal seriously corrupt data (3), extensive slightly corrupt data (5), extensive seriously corrupt data (7), all data totally corrupt (9)

- Loss of availability: How much service could be lost and how vital is it? Minimal secondary services interrupted (1), minimal primary services interrupted (5), extensive secondary services interrupted (5), extensive primary services interrupted (7), all services completely lost (9)
- Loss of accountability: Are the threat agents' actions traceable to an individual? Fully traceable (1), possibly traceable (7), completely anonymous (9)

Estimation of business impact

The business impact stems from the technical impact, but its estimation requires a deep understanding of what is important to the company running the application. It is the business impact which justifies investment in security, particularly from an executive standpoint.

- Financial damage: How much financial damage will result from an exploit? Less than the cost to fix the vulnerability (1), minor effect on annual profit (3), significant effect on annual profit (7), bankruptcy (9)
- Reputation damage: Would an exploit result in reputation damage that would harm the business? Minimal damage (1), loss of major accounts (4), loss of goodwill (5), brand damage (9)
- Non-compliance: How much exposure does non-compliance introduce? Minor violation (2), clear violation (5), high-profile violation (7)
- Privacy violation: How much personally identifiable information could be disclosed? One individual (3), hundreds of people (5), thousands of people (7), millions of people (9)

Likelihood and impact are rated as either low, medium, or high, assuming that a mean score less than three is low, three to less than six is medium, and six to nine is high (Figures 2.1 and 2.2). Business impact is used in preference to technical impact in this assessment, unless accurate calculation of business impact is not possible.

The establishment, maintenance, and continuous update of an Information Security Management System (ISMS) provide a strong indication that a company is using a systematic approach for the identification, assessment, and management of information security risks.

Different methodologies have been proposed to manage IT risks, each of them divided into processes and steps.

The *Certified Information Systems Auditor Review Manual 2006* provides the following definition of risk management: "Risk management is the process of identifying vulnerabilities and threats to the information resources used by an organization in achieving business objectives,

Overall risk severity

	High	Medium	High	Critical
Impact	Medium	Low	Medium	High
	Low	Note	Low	Medium
		Low	Medium	High
			Likelihood	

Figure 2.1 Calculate risk using the following table.

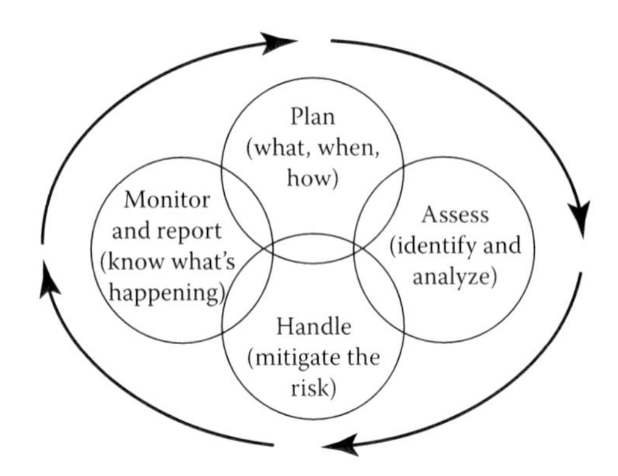

Figure 2.2 IT risk management. (Risk Management Elements. Main article: IT risk management. IT risk management can be considered a component of a wider enterprise risk-management system.)

and deciding what countermeasures, if any, to take in reducing risk to an acceptable level, based on the value of the information resource to the organization."

Risk management is the process that allows IT managers to balance the operational and economic costs of protective measures and achieve gains in mission capability by protecting the IT systems and data that support their organizations' missions. This process is not unique to the IT environment; indeed it pervades decision-making in all areas of our daily lives.

The head of an organizational unit must ensure that the organization has the capabilities needed to accomplish its mission. These mission owners must determine the security capabilities that their IT systems must have to provide the desired level of mission support in the face of real-world threats. Most organizations have tight budgets for IT security; therefore, IT security spending must be reviewed as thoroughly as other management decisions. A well-structured risk-management methodology, when used effectively, can help management identify appropriate controls for providing the mission-essential security capabilities.

Risk management in the IT world is quite a complex, multi-faced activity, related in many ways to other complex activities.

National Information Assurance Training and Education Center defines risk in the IT field as:

- The total process to identify, to control, and minimize the impact of uncertain events. The objective of the risk-management program is to reduce risk and obtain and maintain DAA [designated approving authority] approval. The process facilitates the management of security risks by each level of management throughout the system life cycle. The approval process consists of three elements: risk analysis, certification, and approval.
- An element of managerial science concerned with the identification, measurement, control, and minimization of uncertain events. An effective risk-management program encompasses the following four phases:
  - i. Risk assessment, as derived from an evaluation of threats and vulnerabilities.
  - ii. Management decision.
  - iii. Control implementation.
  - iv. Effectiveness review.
- The total process of identifying, measuring, and minimizing uncertain events affecting AIS [automated information system] resources. It includes risk analysis, cost-benefit analysis, safeguard selection, security test and evaluation, safeguard implementation, and systems review.
- The total process of identifying, controlling, and eliminating or minimizing uncertain events that may affect system resources. It includes risk analysis, cost-benefit analysis, selection, implementation and test, security evaluation of safeguards, and overall security review.

Risk management as part of enterprise risk management

Some organizations have, and many others should have, a comprehensive enterprise risk management (ERM) in place. Four objectives categories are addressed, according to the Committee of Sponsoring Organizations of the Treadway Commission (COSO).

- Strategy: High-level goals, aligned with and supporting the organization's mission
- Operations: Effective and efficient use of resources

- Financial reporting: Reliability of operational and financial reporting
- Compliance: Compliance with applicable laws and regulations

According to the Risk IT framework by ISACA (formerly the Information Systems Audit and Control Association), IT risk is transversal to all four categories. The IT risk should be managed in the framework of enterprise risk management, with the risk appetite and sensitivity of the whole enterprise guiding the IT risk-management process. ERM provides the context and business objectives to IT risk management.

Risk-management methodology

The term methodology means an organized set of principles and rules that drive action in a particular field of knowledge. A methodology does not describe specific methods; nevertheless it does specify several processes that need to be followed.

These processes constitute a generic framework. They may be broken down into sub-processes, they may be combined, or their sequence may change. However, any risk-management exercise must carry out these processes in one form or another.

Due to the probabilistic nature and the need for cost-benefit analysis, the IT risks are managed following a process that can be divided in the following steps per NIST SP 800-30:

- Risk assessment
- Risk mitigation
- Evaluation and assessment

Effective risk management must be totally integrated into the systems development life cycle.

Information risk analysis conducted on applications, computer installations, networks, and systems under development should be undertaken using structured methodologies.

Security is the degree of resistance to, or protection from, harm. The term applies to any vulnerable and valuable asset, such as a person, dwelling, community, nation, or organization. Security provides a form of protection, where a separation is created between the assets and the threat. These separations are generically called "controls," and sometimes include changes to the asset or the threat.

Perceived security compared to real security

Perceptions of security may correlate poorly with measureable objective security. For example, the fear of earthquakes has been reported to be more common than the fear of slipping on the bathroom floor, although the latter kills many more people than the former.

Similarly, the perceived effectiveness of security measures is sometimes different from the actual security provided by the measures. The presence of security protection may be mistaken for security itself. For example, although two computer security programs could be interfering with each other and even cancelling each other's effect, the computer's owner may believe they are getting double the protection.

Security theater is a derogative term for the deployment of measures primarily aimed at raising subjective security without a genuine or commensurate concern for their effects on objective security. For example, some consider the screening of airline passengers based on static databases to have been a form of security theater and that the Computer-Assisted Passenger Prescreening System has actually *decreased* objective security.

Perceived security can increase objective security when it affects or deters malicious behavior. This may be the case with highly visible security protections, such as video surveillance, home alarm systems, or vehicle anti-theft systems such as a vehicle tracking system or warning sign. Since some intruders will decide not to attempt to break into such areas or vehicles, there may be less damage to windows, in addition to protection of valuable objects inside. Without such advertisement, an intruder might, for example, approach a car, break the window, and then flee in response to an alarm being triggered. In either case, neither the car itself nor the objects inside would be stolen, but with the presence of perceived security the risk of the car's windows being damaged is diminished.

Categorizing security

The body of literature on the analysis and categorization of security is immense. Part of the reason for this is that, in most security systems, the "weakest link in the chain" is the most important. The situation is asymmetric since the "defender" must cover all points of attack, while the attacker need only identify a single weak point upon which to concentrate.

Security categorization

IT realm	**Physical realm**	**Political**
• Application security • Computing security • Data security • Information security • Network security	• Airport security • Food security • Home security • Infrastructure security • Physical security • Port security/Supply chain security • School security • Shopping center security	• Homeland security • Human security • International security • National security • Public security **Monetary** • Financial security

Aviation security is a combination of material and human resources and measures intended to counter unlawful interference with aviation.

Operations security (OPSEC) complements other "traditional" security measures that evaluate the organization from an adversarial perspective.

Security concepts

Certain concepts recur throughout different fields of security:

- *Assurance* is the level of guarantee that a security system will behave as expected
- A *countermeasure* is a way to stop a threat from triggering a risk event
- *Defense in depth*—never rely on one single security measure alone
- A *risk* is a possible event which could cause a loss
- A *threat* is a method of triggering a risk event that is dangerous
- *Vulnerability* is a weakness in a target that can potentially be exploited by a security threat
- *Exploit*—a vulnerability that has been triggered by a threat—a risk of 1.0 (100%)

Security policy is a definition of what it means to *be secure* for a system, organization, or other entity. For an organization, it addresses the constraints on behavior of its members as well as constraints imposed on adversaries by mechanisms such as doors, locks, keys, and walls. For systems, the security policy addresses constraints: on functions and flow within them, on access by external systems and adversaries including programs, and on access to data by people.

Significance

If it is important to be secure, then it is important ensure the robustness of the mechanisms of security policy. There are many organized methodologies and risk-assessment strategies to assure security policies' enforcement and completeness. In complex systems, such as information systems, policies can be decomposed to facilitate the allocation of security mechanisms to enforce sub-policies. However, this practice has pitfalls. It is too easy to simply go directly to the sub-policies, which are essentially the rules of operation, and dispense with the top-level policy. This gives the false sense that the rules of operation address some overall definition of security when they do not. Because it is so difficult to think clearly and comprehensively about security, rules of operation stated as "sub-policies" with no "super-policy" usually turn out to be rambling rules that fail to enforce anything with completeness. Consequently, a top-level security policy is essential to any serious security scheme, and sub-policies and rules of operation are meaningless without it.

Conclusion

Evaluations of security concerns become obsolete as technology progresses and new threats and vulnerabilities arise. Continuous security evaluation of organizational products, services, methods, and technology is essential to maintain effective overall security measures within one's organization. Security concerns which have previously been evaluated need to be re-evaluated. A continuous security evaluation mechanism within the organization is critical to achieving the overall security objectives. The re-evaluation process is tied to the dynamic security-requirement management process discussed above.

chapter three

Threat/vulnerability assessments and risk analysis

Introduction

There are *fundamental principles* underlying the threat-assessment approach. The *first principle* is that targeted violence is the result of an understandable and often discernible process of thinking and behavior. Acts of targeted violence are neither impulsive nor spontaneous. Ideas about monitoring an attack usually develop over a considerable period of time. In targeted violence, the subject must engage in planning around a series of critical factors such as which target(s) to select, the proper time and approach, and the means for violence.

A potential attacker may collect information about the target, the setting of the attack, or about similar attacks. He or she may communicate ideas to others. For some of these individuals the process of planning and thinking about the attack dominates their lives and provides a sense of purpose or an attainable goal by which they see an end to their emotional pain.

The *second principle* is that violence stems from an interaction between the potential attacker, past stressful events, a current situation, and the target. As noted above in the discussion of the risk-assessment model, researchers and practitioners are moving away from exclusive focus on the individual and toward a more situational/contextual understanding of risk.

An assessment of the attacker may consider relevant risk factors, development and evolution of ideas concerning the attack, preparatory behaviors, and an appraisal of how the individual has dealt with unbearable stress in the past. When usual coping mechanisms are ineffective, people often react by becoming physically ill, psychotic, self-destructive, or violent toward others. It is useful to consider how the potential attacker has responded in the past when stressful events overwhelmed his/her coping resources.

An assessment of the risk may be informed by an examination of the person's history of response to traumatic major changes or losses, such as loss of a loved one (e.g., ending of an intimate relationship or loss of a parent) or loss of status (e.g., public humiliation, failure, or rejection, or loss of job or financial status). The salience of the risk may be determined by examining the types of event that have led the individual to experience life as unbearably stressful, the response to those events, and the likelihood that they may recur.

In addition to assessing the potential attacker and past stressful events, the evaluator must also appraise the current situation and the target. Consideration of the current situation includes both an appraisal of the likelihood that past life events which triggered consideration of self-destructive or violent behavior will recur (or are recurring) and an assessment of how others in the subject's environment are responding to his/her perceived stress and potential risk.

Since others may act to prevent violence, it is useful to know whether people around the subject support, accept, or ignore the threat of violence, or whether they express disapproval and communicate that violence is an impermissible and unacceptable solution to the problem. Finally, an evaluator must assess relevant factors about the intended target, including the subject's degree of familiarity with the target's work and lifestyle patterns, the target's vulnerability, and the target's sophistication about the need for caution.

The *third principle* is that a key to investigation and resolution of threats is that those who commit acts of targeted violence often engage in discrete behaviors that precede and are linked to their attacks, including thinking, planning, and logistical preparations.

Attack-related behaviors may move along a continuum beginning with the development of an idea about an attack, moving to communication of these ideas or an inappropriate interest in others, to following, approaching, and visiting the target or scene of the attack, even with lethal means. Learning about and analyzing these behaviors may be critical to an appraisal of risk.

Threat assessment

The first step in a risk-management program is a threat assessment. A threat assessment considers the full spectrum of threats (i.e., natural, criminal, terrorist, accidental, etc.) for a given facility/location. The assessment should examine supporting information to evaluate the likelihood of occurrence for each threat. For natural threats, historical data concerning frequency of occurrence for given natural disasters such as tornadoes, hurricanes, floods, fire, or earthquakes can be used to determine the credibility of the given threat.

For criminal threats, the crime rates in the surrounding area provide a good indicator of the type of criminal activity that may threaten the facility. In addition, the type of assets and/or activity located in the facility may also increase the target's attractiveness in the eyes of the aggressor. The type of assets and/or activity located in the facility will also relate directly to the likelihood of various types of accidents. For example, a facility that utilizes heavy industrial machinery will be at higher risk for serious or life-threatening job-related accidents than a typical office building.

Vulnerability assessment

Once the credible threats are identified, a vulnerability assessment must be performed. The vulnerability assessment considers the potential loss from a successful attack as well as the vulnerability of the facility/location to an attack. Impact of loss is the degree to which the mission of the agency is impaired by a successful attack from the given threat.

A key component of the vulnerability assessment is properly defining the ratings for impact of loss and vulnerability. These definitions may vary greatly from facility to facility. For example, the amount of time that mission capability is impaired is an important part of impact of loss. If the facility being assessed is an air route traffic control tower, a downtime of a few minutes may be a serious impact of loss, while for a social security office a downtime of a few minutes would be have a minor impact. A sample set of definitions for impact of loss is provided below. These definitions are for an organization that generates revenue by serving the public.

- *Devastating*: The facility is damaged and contaminated beyond habitable use. Most items/assets are lost, destroyed, or damaged beyond repair/restoration. The number of visitors to other facilities in the organization may be reduced by up to 75% for a limited period of time.
- *Severe*: The facility is partially damaged and contaminated. Examples include partial structure breach resulting in weather/water, smoke, impact, or fire damage to some areas. Some items/assets in the facility are damaged beyond repair, but the facility remains mostly intact. The entire facility may be closed for a period of up to two weeks and a portion of the facility may be closed for an extended period of time (more than one month). Some assets may need to be moved to remote locations to protect them from environmental damage. The number of visitors to the facility and others in the organization may be reduced by up to 50% for a limited period of time.
- *Noticeable*: The facility is temporarily closed or unable to operate, but can continue without an interruption of more than one day. A limited number of assets may be damaged, but the majority of the facility is not affected.

- The number of visitors to the facility and others in the organization may be reduced by up to 25% for a limited period of time.
- *Minor*: The facility experiences no significant impact on operations (downtime is less than four hours) and there is no loss of major assets.

Vulnerability is defined to be a combination of the attractiveness of a facility as a target and the level of deterrence and/or defense provided by the existing countermeasures. *Target attractiveness* is a measure of the asset or facility in the eyes of an aggressor and is influenced by the function and/or symbolic importance of the facility. Sample definitions for vulnerability ratings follow.

- *Very high*: This is a high-profile facility that provides a very attractive target for potential adversaries, and the level of deterrence and/or defense provided by the existing countermeasures is inadequate.
- *High*: This is a high-profile regional facility or a moderate-profile national facility that provides an attractive target and/or the level of deterrence and/or defense provided by the existing countermeasures is inadequate.
- *Moderate*: This is a moderate-profile facility (not well known outside the local area or region) that provides a potential target and/or the level of deterrence and/or defense provided by the existing countermeasures is marginally adequate.
- *Low*: This is not a high-profile facility and provides a possible target and/or the level of deterrence and/or defense provided by the existing countermeasures is adequate.

The vulnerability assessment may also include detailed analysis of the potential impact of loss from an explosive, chemical, or biological attack. Professionals with specific training and experience in these areas are required to perform these detailed analyses. A sample of the type of output that can be generated by a detailed explosive analysis is shown in Figure 3.1. This graphic representation of the potential damage to a facility from an explosive attack allows a building owner to quickly interpret the results of the analysis, although a more fully detailed and quantitative engineering response would be required to design a retrofit upgrade. In addition, similar representations can be used to depict the response of an upgraded facility to the same explosive threat.

This allows a building owner to interpret the potential benefit that can be achieved by implementing various structural upgrades to the building frame, wall, roof, and/or windows (Figure 3.2).

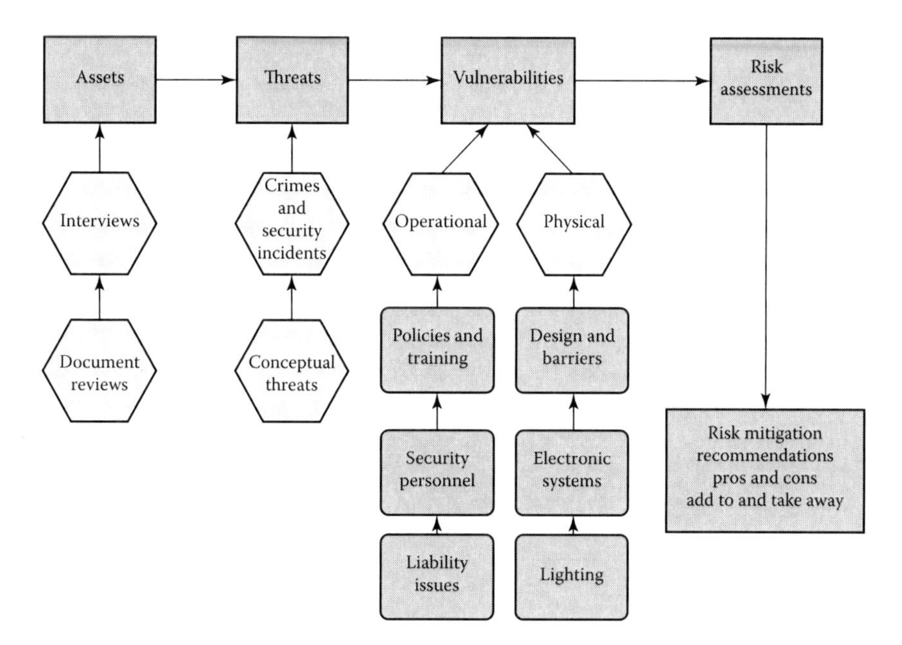

Figure 3.1 The risk-assessment process.

Risk analysis

A combination of the impact of loss rating and the vulnerability rating can be used to evaluate the potential risk to the facility from a given threat.

Flowchart depicting the basic risk-assessment process (Figure 3.1): this is just one of many types of flowcharts, the type used will vary depending on one's needs.

Upgrade recommendations

Based on the findings from the risk analysis, the next step in the process is to identify countermeasure upgrades that will lower the various levels of risk.

If minimum standard countermeasures for a given facility level are not currently present, these countermeasures should automatically be included in the upgrade recommendations. Additional countermeasure upgrades above the minimum standards should be recommended as necessary to address the specific threats identified for the facility.

Flowchart depicting the basic risk-assessment process; this is just one of many types of flow charts. This is just an example, and it will vary depending on one's needs (Figure 3.2).

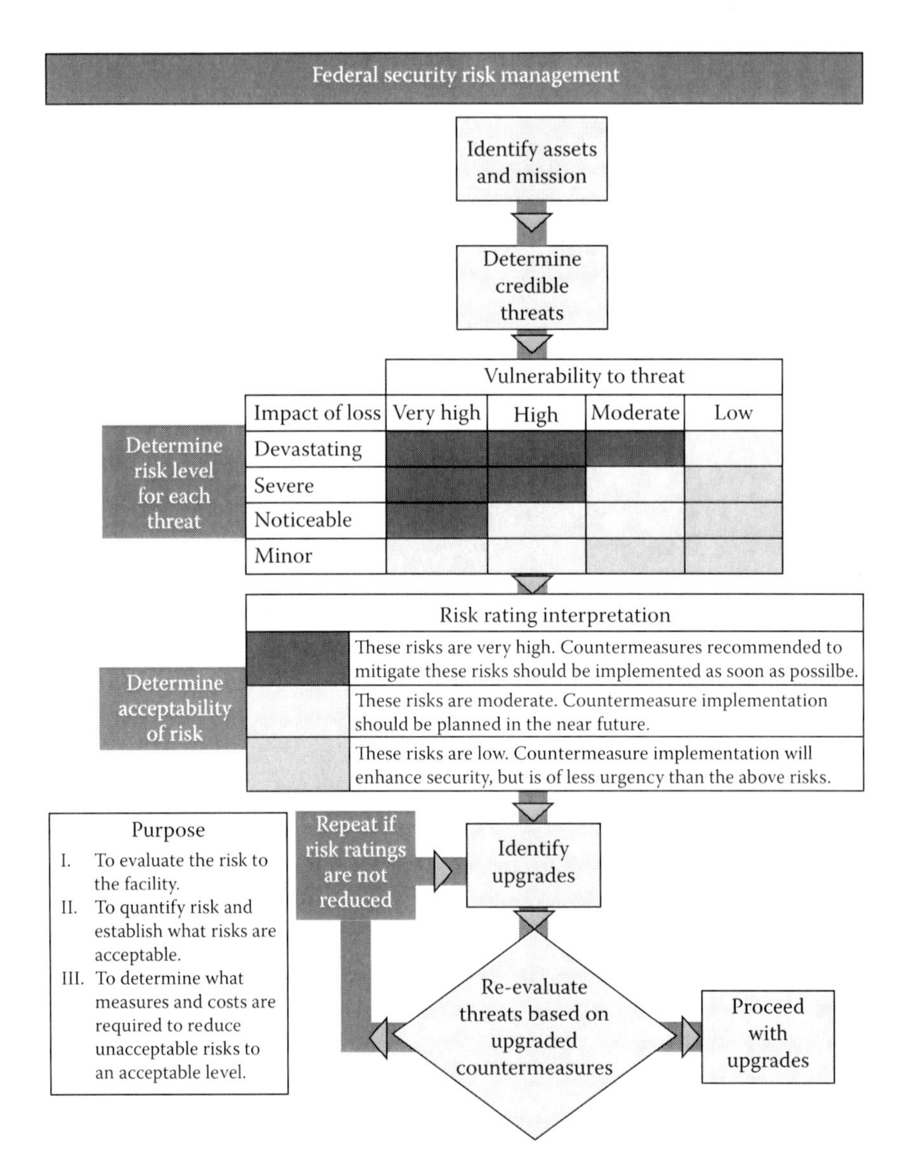

Figure 3.2 Security risk management.

The estimated capital cost of implementing the recommended countermeasures is usually provided in the threat/vulnerability assessment report. The estimated installation and operating costs for the recommended countermeasures are also usually provided in the threat/vulnerability assessment report. All operating costs are customarily estimated on a per year basis.

Re-evaluation of risks

The implementation of the recommended security and/or structural upgrades should have a positive effect on the impact of loss and/or the vulnerability ratings for each threat. The final step in the process is to re-evaluate these two ratings for each threat in light of the recommended upgrades. Using an exterior explosive threat as an example, the installation of window retrofits (i.e., security window film, laminated glass, etc.) will not prevent the explosive attack from occurring, but it should reduce the impact of loss/injury caused by hazardous flying glass. Therefore, the impact of loss rating for an explosive threat would improve, but the vulnerability rating would stay the same.

Understanding threats

As responders we must understand threat components, which we must then counter. The threat has three components: aggressors; their tactics; and their associated weapons, explosives, and tools.

There are four types of aggressors that engineers must understand and plan against in a low-intensity conflict environment (Figure 3.3).

Criminals are subdivided into three categories: unsophisticated, sophisticated, and organized. Unsophisticated criminals are unskilled in the use of weapons and tools and have no formal organization. Their targets are those that meet their immediate needs, such as drugs, money, and pilferable items. They are interested in targets that pose little risk. Sophisticated criminals working singly are organized and efficient in the

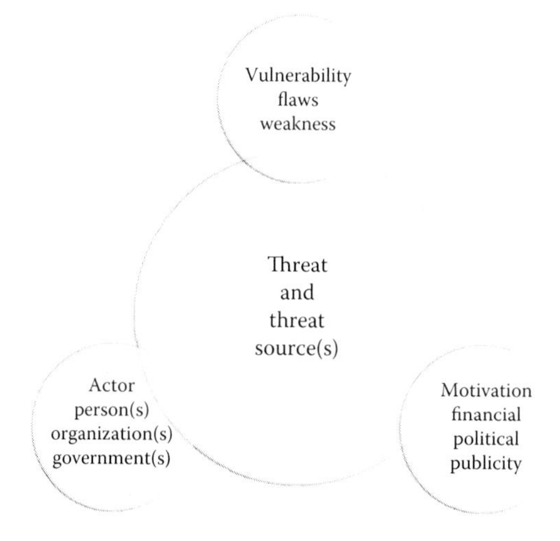

Figure 3.3 Threat assessment types.

use of certain weapons and tools. They target high-value assets and frequently steal large quantities. Organized criminal groups are sophisticated and rely on specialists to obtain equipment to achieve specific goals. Targets of organized criminal groups may include large quantities of money, equipment, arms, ammunition, and explosives.

Protestors may be categorized as vandals, activists, or extremists. Engineers must be concerned about all violent protestors. Protestors are politically or issue-orientated and act out of frustration, discontent, or anger. Their primary objectives include destruction and publicity.

Vandals and activists are unsophisticated and superficially destructive. They generally do not intend to injure people. Extremist groups are moderately sophisticated and more destructive. Their actions are frequently overt and may involve individuals as targets. Extremists, or terrorists are motivated by an ideology, a political cause, or an issue.

Terrorists commonly work in small, well-organized groups. They are sophisticated and possess efficient planning capabilities. Terrorist objectives include death, destruction, theft, and publicity. Terrorist groups are generally classified by their government affiliation.

Subversives are classified into two groups: saboteurs and spies. Saboteurs include guerrillas and commandos. They are very sophisticated and highly skilled and employ meticulous planning. Saboteurs commonly operate in small groups and have an unlimited arsenal. Their objectives include death and destruction. They often target mission-critical personnel, equipment, or operations. Spies are highly skilled and very sophisticated. They are generally foreign agents but frequently employ insiders. They target military information and attempt to avoid detection. In some cases, they may employ activists or other aggressors.

Individuals posing a threat employ a wide range of tactics to accomplish their objectives. These strategies have been categorized into 15 tactics, which are specific methods of achieving an aggressor's goals. The following descriptions of aggressor tactics will assist engineer planners in developing protective methods, devices, facilities, and systems.

Moving-vehicle bomb: used when an aggressor's goal is to damage or destroy a facility (or assets within a facility) or to kill people within the blast area. The moving-vehicle bomb is used in a suicide attack where an explosive-laden vehicle is driven into a facility and detonated.

Stationary-vehicle bomb: used when an aggressor's primary objective is to damage or destroy a facility (or assets within a facility). This type of bomb maybe detonated by time delay or remote control. This attack has three versions:

1. An explosive-laden vehicle is driven to a preselected location and abandoned.

2. Explosives are placed in an unsuspecting person's car. He then unknowingly delivers the bomb to the targeted facility.
3. Someone is coerced into delivering a vehicle bomb.

Exterior attack: used when an aggressor's goal is to damage or destroy a facility (or assets within a facility) and kill or injure its occupants. This attack is conducted at close range to a facility or exposed asset. Using clubs, rocks, improvised incendiary devices, hand grenades, or hand-placed bombs, the aggressor attempts to inflict destruction and death.

Standoff weapons attack: used when an aggressor's goal is to damage or destroy a facility (or assets within a facility) and kill or injure its occupants. These attacks are executed using military or improvised direct- and indirect-fire weapons, such as antitank weapons and mortars.

Ballistic attack: used when an aggressor's goal is to kill or injure a facility's occupants. Using small arms at varying distances, the aggressor attempts to inflict death.

Forced entry: used when an aggressor's goals are to steal or destroy assets, compromise information, or disrupt operations. Using small arms or forced-entry tools, the aggressor enters a facility through an existing passage or creates a new opening in the facility.

Covert entry: used when an aggressor's goals are identical to those listed for the forced-entry tactic. The difference in these entries is that the aggressor will attempt to enter the facility covertly using false credentials. The aggressor may attempt to carry weapons or explosives into the facility.

Insider compromise: used when an aggressor's goals are similar to those listed for the forced-entry tactic. The aggressor uses an insider (one who has legitimate access to a facility) to accomplish their prescribed objectives.

Electronic eavesdropping: used by an aggressor to monitor electronic emanations from computers, communications, and related equipment. This eavesdropping is normally done from outside a facility or restricted area.

Acoustical eavesdropping: an aggressor uses a listening device to monitor voice communication and other audible information.

Visual surveillance: used by aggressors employing ocular and photographic devices to monitor a facility, installation, or mission operations.

Mail bombs: used when the aggressor's objective is to kill or injure people. Small bombs or incendiary devices are incorporated into envelopes or packages that are delivered to the targeted individual.

Supplies bombs: used when the aggressor's objective is to kill or injure people or destroy facilities. Bombs or incendiary devices, generally larger than those found in mail bombs, are incorporated into various containers and delivered to facilities or installations.

Airborne contamination: used when the aggressor's objective is to kill people. The aggressor uses chemical or biological agents to contaminate the air supply of a facility or installation.

Waterborne contamination: used when an aggressor's objective is to kill people. The aggressor uses chemical, biological, or radiological agents to contaminate the water supply of a facility or installation.

Aggressors use various types of weapons, explosives, and tools to attain their objectives. Weapons range from clubs and rocks to mortars. Explosives are commonly used to destroy facilities and housing assets and to kill people.

Tools are primarily used in forced-entry operations to breach protective components or barriers. Understanding the aggressor's options will aid the engineer in protecting forces from these items. Listed below are various weapons, explosives, and tools and their potential uses.

Rocks and clubs: used in exterior building attacks to damage exterior building components or exposed assets or to injure people.

Incendiary devices: used to damage the facility's exterior or sabotage other assets. These include hand-held torches and improvised incendiary devices (IID). An example is a "Molotov cocktail."

Firearms: used in a ballistic tactic to attack facility assets from a distance and in the forced-entry tactic to overpower guards. These include pistols, rifles, shotguns, and submachine guns, both military and civilian. Weapons capabilities are outlined in the Security Engineering Manual.

Antitank weapons and mortars: used in standoff attacks on facilities. For example, the direct-fire antitank weapons most often used by terrorists are the Soviet, rocket-propelled grenade RPG-7 and the U.S. light antitank weapon (LAW). These weapons increase the terrorist's ability to penetrate and damage a facility and to kill or injure people. Mortars are indirect-fire weapons and include both military and improvised versions.

Nuclear, biological, and chemical: delivered as airborne or waterborne gases, liquids, aerosols, or solids. Very powerful chemical agents can be manufactured with relative ease from commercially available products. Biological agents can be grown in unsophisticated home laboratories. Radiological agents are radioactive elements that pose a potential threat to water supplies. They can be delivered in liquid or solid form.

Improvised explosive devices (IED): used in the exterior attack, mail and supplies bomb deliveries, forced-entry, covert-entry, and insider-compromise tactics to destroy assets and to injure or kill people. They are commonly "homemade" bombs made of plastic explosives or trinitrotoluene (TNT). Plastic explosives are chosen by terrorist and extremist protestor groups because they are easily molded, stable, and difficult to detect.

Hand grenades: used in exterior attacks to injure or kill people. These include common military antipersonnel and fragmentation hand grenades.

Vehicle bombs: used to destroy facilities and kill people. They contain large quantities of explosives and have the potential to do catastrophic damage.

Potential aggressors have access to a wide variety of tools, ranging from forced-entry tools, (hand and power tools, cutting torches, and burn bars) to sophisticated surveillance tools and devices. The quality and effectiveness of tools and devices used depends on the type of aggressor. The more sophisticated, trained, and organized the aggressor is, the more dangerous his tools and devices will be.

Electronic risk assessment and vulnerabilities: Computer/network internal and external threats

The purpose of a threat risk assessment is to categorize enterprise assets, examine the different threats that may jeopardize them, and identify and correct the most immediate and obvious security concerns.

While taking tactical measures to correct immediate problems is important, understanding the different threats and risks will enable management to make informed decisions about security so they can apply appropriate, cost-effective safeguards in the longer term. This balanced approach allows for strategic positioning by logically applying risk-mitigation strategies in a controlled and economical manner rather than informal, and often expensive, implementations.

The threat, risk, and vulnerability assessment is an objective evaluation of threats, risks, and vulnerabilities in which assumptions and uncertainties are clearly considered and presented. Part of the difficulty of risk management is that both of the quantities with which risk assessment is concerned—potential loss and probability of occurrence—can be very difficult to measure. The chance of error in the measurement of these two concepts is large.

A risk with a large potential loss and a low probability of occurring is often treated differently from one with a low potential loss and a high likelihood of occurring. While in theory they may be of nearly equal priority, in practice it can be very difficult to prioritize both when faced with the scarcity of resources, especially time, in which to conduct the risk-management process.

One of the challenges in computer security is deciding how much security is necessary for proper control of system and network assets. This decision depends on conducting a threat assessment or, more specifically; what do you have and who would want it? While this is relatively simple to state, assessing a corporate network threat is not easy without a structured approach.

IT security risk/threat management assesses information security policies, processes, and technologies to identify weaknesses, categorize security risks, and recommend improvements. Security assessment and risk analysis helps fortify the environment and improves compliance with industry regulations by providing a comprehensive assessment of each important aspect of the security program including:

- Internal and external controls
- Policies and procedures
- Gaps versus regulations and best practices
- Vulnerabilities and threats

IT security risk assessment and analysis: Reporting objectives

The report objectives of the security assessment and risk analysis are to provide management with clear and concise answers to the following questions:

- Within the scope of the control areas being tested, how well are information-based assets protected from internal and external threats?
- Are management, administrative, physical, and technical- and policy-based controls adequate?
- How do the controls compare to others in the industry?
- What is the quickest, most cost-effective way to manage risk to an acceptable level?

Methodology

Best-practice benchmarks may be used to identify select control gaps and strengths. A gap analysis-based approach allows a company sufficient control-visibility to set objectives and priorities for remediation efforts. It also allows documentation and representation of current control activities to regulatory auditors and examiners in the best context possible, as a best practice.

The threat vulnerability assessment tool

The threat vulnerability assessment requires organizations to conduct a risk vulnerability and threat assessment. The process concludes with a security overall assessment.

Sample risk assessment looking for the following issues and threats.

Risk analysis description

It is significant to note that a risk analysis, as defined by this methodology, is not an audit or an evaluation carried out by one entity upon another.

It is, rather, a collaborative and objective fact-finding exercise between a managerial and technical team trained in this methodology and a managerial and technical team responsible for the information system(s) being analyzed. All findings included in the analysis must be substantiated by an objective reference (e.g., vulnerability scans, physical security reviews, incident reports, log records, etc.), which is then included in the appendices to the final risk analysis report. The output of this process helps to identify appropriate controls for reducing or eliminating risk during the risk mitigation process.

Threat and vulnerability analysis

The mission of the threat and vulnerability analysis team is to provide a detailed analysis, following the risk analysis methodology, of the specific threats and vulnerabilities associated with an IT system's environment and configuration. The goal of this exercise is to develop an objective list of system vulnerabilities (flaws or weaknesses) that could be exploited by potential threat sources.

Threat and vulnerability analysis description

In addition to research and interviews, automated vulnerability scanning tools are used to identify system vulnerabilities efficiently. Frequently, vulnerabilities can be directly linked to missing or incomplete system-specific IT controls, or more broadly focused infrastructure-level common IT controls. The detailed analysis output from the "threat and vulnerability" step is the foundation upon which the next step in the risk analysis process is built.

Attacks and threats

- Handling destructive malware
- Understanding denial-of-service attacks
- Understanding hidden threats: corrupted software files
- Understanding hidden threats: rootkits and botnets
- Recognizing fake anti-viruses
- Avoiding the pitfalls of online trading
- Avoiding social engineering and phishing attacks
- Dealing with cyber bullies
- Identifying hoaxes and urban legends
- Preventing and responding to identity theft
- Recognizing and avoiding spyware
- Recovering from viruses, worms, and Trojan horses
- A working plan can be used to conduct the threat and vulnerability assessment as well as to define the components of the process including:

- Administrative safeguards
- Logical safeguards
- Physical safeguards
- Demographics of each physical location
- Access to each facility at each physical location
- Environmental factors associated with each physical location
- IT and business processes at each physical location
- A risk ranking matrix with a scoring mechanism that looks at vulnerability, as measured by probability of the threat occurring versus the impact of the loss and rules for scoring the risk
- A risk conveys the risk assessor's judgment as to the nature and presence or absence of risks, along with information about how the risk was assessed, where assumptions and uncertainties still exist, and where policy choices will need to be made. Risk characterization takes place in both human-health and ecological risk assessments.
- Each component of the risk assessment (e.g., hazard assessment, dose-response assessment, exposure assessment) has an individual risk, assumptions, limitations, and uncertainties. The set of these individual risk characterizations provides the information basis to write an integrative risk-characterization analysis.
- The final, overall risk consists of the individual risk characterizations plus an integrative analysis. The overall risk characterization informs the risk manager and others about the approach to conducting the risk assessment.

Principles of conducting risk characterizations

A good risk characterization will restate the scope of the assessment, express results clearly, articulate major assumptions and uncertainties, identify reasonable alternative interpretations, and separate conclusions from policy judgments. Risk characterization policy calls for conducting risk characterizations in a manner that is consistent with the following principles.

- Transparency: The characterization should fully and explicitly disclose the risk assessment methods, default assumptions, logic, rationale, extrapolations, uncertainties, and overall strength of each step in the assessment.
- Clarity: The products from the risk assessment should be readily understood by readers inside and outside of the risk-assessment process. Documents should be concise, free of jargon, and should use understandable tables, graphs, and equations as needed.
- Consistent: The risk assessment should be conducted and presented in a manner which is consistent with Environmental Protection

Agency (EPA) policy, and consistent with other risk characterizations of similar scope prepared across programs within the EPA.
- Reasonable: The risk assessment should be based on sound judgment, with methods and assumptions consistent with the current state-of-the-science and conveyed in a manner that is complete, balanced, and informative.

In the field of risk assessment, characterizing the nature and magnitude of human-health or environmental risks is arguably the most important step in the analytical process. In this step, data on the dose-response relationship between an agent and outcomes are integrated with estimates of the degree of exposure of a population to characterize the likelihood and severity of risk. Although the purpose of risk characterizations is to make sense of the available data and describe what they mean to a broad audience, this step is often given insufficient attention in health risk evaluations. Too often, characterizations fail to interpret or summarize risk information in a meaningful way, or they present single numerical estimates of risk without an adequate discussion of the uncertainties inherent in key exposure parameters or the dose-response assessment, model assumptions, or analytical limitations. Consequently, many users of risk information have misinterpreted the findings of a risk assessment or have false impressions about the degree of accuracy (or the confidence of the scientist) in reported risk estimates.

In this article we collected and integrated the published literature on conducting and reporting risk characterizations to provide a broad, yet comprehensive, analysis of the risk characterization process as practiced in the United States and some other countries. Specifically, the following eight topics are addressed:

1. Objective of risk characterization
2. Guidance documents on risk characterization
3. Key components of risk characterizations
4. Toxicity criteria for evaluating health risks
5. Descriptors used to characterize health risks
6. Methods for quantifying human-health risks
7. Key uncertainties in risk characterizations
8. The risk decision-making process

The risk characterization guide is designed to provide risk assessors, risk managers, and other decision-makers an understanding of the goals and principles of risk characterization, the importance of planning and scoping for a risk assessment, the essential elements to address in a risk characterization, the factors that are considered in decision-making by risk managers, and the forms that risk characterization takes for different audiences.

Conclusion

A risk conveys the risk assessor's judgment as to the nature and presence or absence of risks, along with information about how the risk was assessed, where assumptions and uncertainties still exist, and where policy choices will need to be made. Risk characterization takes place in both human-health risk and ecological risk assessments.

Each component of the risk assessment (e.g., hazard assessment, dose-response assessment, exposure assessment) has an individual risk, assumptions, limitations, and uncertainties. The set of these individual risk characterizations provides the information basis to write an integrative risk-characterization analysis.

The final, overall risk consists of the individual risk characterizations plus an integrative analysis. The overall risk characterization informs the risk manager and others about the approach to conducting the risk assessment.

chapter four

Defense support of civil authorities (DSCA)

Defense support of civil authorities

Defense Support of Civil Authorities (DSCA) is the process by which United States' military assets and personnel can be used to assist in missions normally carried out by civil authorities. These missions have included responses to natural and man-made disasters, law enforcement support, special events, and other domestic activities. A recent example of the use of DSCA is the military response to Hurricane Katrina.

DSCA is the overarching guidance as to how interventions by the United States military can be requested by a federal agency and the procedures that govern the actions of the military during employment.

The Directorate of Military Support for Domestic Operations (DOMS) is the functional process manager of DSCA, and is located inside each state's Joint Operation Center (JOC). The normal course of action is for the Office of Emergency Management within the state to request military support through the JOC. In turn, the JOC under the authority of the DOMS will initiate military equipment deployment to the Federal Emergency Management Agency's (FEMA) emergency support functions.

The provision of DSCA is codified in Department of Defense Directive 3025.18. This Directive defines DSCA as:

- Support provided by U.S. federal military forces, National Guard, DOD civilians, DOD contract personnel, and DOD component assets, in response to requests or assistance from civil authorities for special events, domestic emergencies, designated law enforcement support, and other domestic activities.
- Support provided by National Guard forces performing duty in accordance with Reference (m) is considered DSCA, but is conducted as a State-directed action also known as civil support.

There are numerous other directives, policies, and laws that shape the military's role in conducting operations in support of other federal agencies.

These include the Insurrection Act, Homeland Security Act, Stafford Act, Economy Act, and Homeland Security Presidential Directive 5. Each of these affects the way the military responds to a request for assistance from an interagency partner.

Once federal forces, deployed in support of DSCA, enter the incident area they come under the operational control of U.S. Northern Command. U.S. Northern Command only controls federal forces deployed into the impact area in response to the incident. National Guard forces deployed under the authority of the Governor remain under control of the Governor.

Activating the DSCA process and providing support to civil authorities cannot impair the ability of the military to conduct its primary mission. It is also critical to understand that the military is always in a supporting role and is never the lead. According to the 2011 DSCA Interagency Partner Guide, a request to the DSCA is made when a disaster, crisis, or special incident occurs and local, tribal, or state authorities can no longer manage the situation. All incidents are controlled at the lowest civilian levels, with the military filling in critical roles. Only federal agencies can request Department of Defense assistance, and only in response to a state need.

All federal agencies can request military assistance by using a simple memo format that contains specific information on what capability is needed and also gives cost reimbursement guidance. Military assets conducting support stay under the control of the military chain of command. Assistance is coordinated with the local responders in the disaster area to ensure the military support is being properly utilized as per the approved request.

If there is a need to change the original mission of assisting forces, the request process starts all over again. When the military has completed the requested support function is determined collaboratively between DOD officials, local government, and federal agencies.

The costs incurred in providing military support must be reimbursed by the agency that requested it. The military's budget does not provide for DSCA support, so it must be paid back in order to maintain the ability to conduct its primary mission.

On the civilian side, the Federal Emergency Management Agency, better known as FEMA, is the main federal responder when a disaster overwhelms a state(s). When a state exhausts all of its resources or is lacking a unique capability during a disaster, it will turn to the federal government for assistance. FEMA will usually be the agency that responds and coordinates the overall federal effort. FEMA has a cornucopia of options available to help a state in need, one of which is to turn to the military. FEMA has the ability to direct most other federal resources under its statutory authority, but it cannot direct the military; rather, it must request the assistance. Military support is a last resort; all other local, state, and federal resources must be exhausted, or a unique requirement that cannot be satisfied by the civilian or federal system must exist prior to the provision of military support.

Requesting the military to respond to a disaster, man-made or natural, is done through a formal process established between FEMA and the Department of Defense. While this process is quite straightforward, it has many integrated steps that require involvement from numerous sources, both military and civilian.

An early event that had a major influence on shaping how the military responds was the 1794 Whiskey Rebellion. It set the stage for establishing the fundamental principles codified in the United States' current laws. Because of the excise tax on whiskey, taxpayers revolted against the federal government. Violence against tax collectors grew to such a level that it prompted presidential intervention. From August to November 1794, federal troops were deployed to Western Pennsylvania as a show of force. Throughout this threat to federal authority, President Washington's guidance was that the military was to support the local civil authorities, not impede them or control them in any way. This underlying principle remains imbedded in the present laws, systems, and processes of how the military interacts within the DSCA environment.

Another important factor governing the actions of the military in executing DSCA is the *Posse Comitatus* Act (PCA). This law was established after the Civil War and prohibited active-duty military (Title 10) from enforcing civil law and order unless directed by the President. The most recent event that caused the President to employ the federal military in direct support of local law enforcement was the Los Angeles riots of 1992, when troops where brought into to help quell the violence. The PCA does not apply to the United States National Guard in their state-support role.

Military aid to the civil power (MACP) (sometimes to the civil authorities) is a term used to describe the use of the armed forces in support of the civil authorities of a state. The term is used in many countries, with slightly different definitions and implications in each.

The *Posse Comitatus* Act, passed in 1878, generally prohibits federal military personnel (except the United States Coast Guard) and units of the United States National Guard under federal authority from acting in a law enforcement capacity within the United States, except where expressly authorized by the Constitution or Congress.

The original act only referred to the Army, but the Air Force was added in 1956, and the Navy and Marine Corps have been included by a regulation of the Department of Defense. This law is mentioned whenever it appears that the Department of Defense is interfering in domestic disturbances. However, the National Guard may still be used for police-like duties once it remains under control of the state, as with the 1967 Detroit riot. Repeated caveats have been added to the *Posse Comitatus* Act over the years by subsequent legislation.

Military operations other than war (MOOTW) focus on deterring war, resolving conflict, promoting peace, and supporting civil authorities in response to domestic crises. The phrase and acronym were coined by the United States military during the 1990s, but have since fallen out of use. The UK military has crafted an equivalent or alternate term, peace support operations (PSO). Both MOOTW and PSO encompass peacekeeping, peacemaking, peace enforcement, and peace building.

MOOTW not involving the use or threat of force include humanitarian assistance and disaster relief. Special agreements exist which facilitate fire-support operations within NATO and the ABCA quadripartite working group, which includes American, British, Canadian, and Australian military contingents. Cooperation is organized in advance with NATO standardization agreements (STANAGs) and quadripartite standardization agreements (QSTAGs). Many countries which need disaster support relief have no pre-existing bilateral agreements: action may be required, based on the situation, to establish such agreements.

MOOTW also involves arms control and peacekeeping

The United Nations (UN) recognizes the vulnerability of civilians in armed conflict. Security Council Resolution 1674 (2006) on the protection of civilians in armed conflict focuses international attention on the protection of civilians in UN and other peace operations. Paragraph 16 anticipates that peacekeeping missions are provided with clear guidelines regarding what missions can and should do to achieve protection goals; that the protection of civilians is given priority in decisions about the use of resources; and that protection mandates are implemented.

MOOTW purposes may include deterring potential aggressors, protecting national interests, and supporting United Nations (UN) objectives.

Peacetime and conflict represent two states of the range of military operations.

- Peacetime is a state in which diplomatic, economic, informational, and military powers are operate together to achieve national objectives.
- Conflict is a unique environment in which the military works closely with diplomatic leaders to control hostilities, and the national objective is focused on the goal of returning to peacetime conditions.

Planners are challenged to find ways to resolve or work around unique arrays of inter-related constraints, issues related to budgeting, training, and force structure. The uncertainties which are inherent or implied include the varying political aspects which are likely to affect unanticipated MOOTW.

In United States military doctrine, military operations other than war include the use of military capabilities across a range of operations that fall short of war. Because of political considerations, MOOTW operations normally have more restrictive rules of engagement (ROE) than wartime operations.

Although the MOOTW acronym is new, the concepts are not. The RAND database identifies 846 military operations other than war between 1916 and 1996 in which the U.S. Air Force or its predecessors played a noteworthy role.

Select American deployments

This is a dynamic list and may never be able to satisfy particular standards for completeness. You can help by expanding it with reliably sourced entries.

- 2005: Nias–Simeulue earthquake: Emergency relief and medical assistance.
- 1990–1994: Operation Promote Liberty: Occupation and peacekeeping mission in Panama after the 1989 United States invasion of Panama.
- 1991: Operation Eastern Exit: Noncombatant evacuation operation to evacuate diplomatic staff and civilians, from the United States and 29 other countries, from the U.S. Embassy in Mogadishu, Somalia as the city plunged into near-anarchy during the Somali Civil War.
- 2001–present: Operation Enduring Freedom: Bush Doctrine continuous operation across numerous countries, mainly Afghanistan, Pakistan, Kyrgyzstan and Uzbekistan.
- 2011: Military intervention in Libya: UN-authorized no-fly zone enforcement in defense of rebel factions in Libya.

Selected British deployments

This is a dynamic list and may never be able to satisfy particular standards for completeness. You can help by expanding it with reliably sourced entries.

- 1948–1960 Malayan Emergency.
- 1995 post-Bosnian War, Operation Deliberate Force.

Select Japanese deployments

This is a dynamic list and may never be able to satisfy particular standards for completeness. You can help by expanding it with reliably sourced entries.

- 2003–2009 Iraq War, Operation Enduring Freedom: Ground Self-Defense Forces, water purification near Basra; Air Self-Defense Forces, cargo and personnel transport; Maritime Self-Defense Forces, supply ships servicing the international flotilla .

Select Australian deployments

This is a dynamic list and may never be able to satisfy particular standards for completeness. You can help by expanding it with reliably sourced entries.

- 1947 UN Consular Commission to Indonesia
- 2005 Nias–Simeulue earthquake, Operation Sumatra Assist: Emergency relief and medical assistance.

Current Australian deployments

- UN Assistance Mission in Afghanistan (UNAMA)
- UN Assistance Mission for Iraq (UNAMI)
- UN Peacekeeping Force in Cyprus (UNFICYP)
- UN Truce Supervision Organization (UNTSO)
- UN Integrated Mission in Timor-Leste (UNMIT)UN Mission in the Sudan (UNMIS)
- UN–African Union Mission in Darfur (UNAMID) (Figures 4.1 through 4.6)

Conclusion

Defense support of civil authorities (DSCA) describes support provided by federal military forces, Department of Defense (DOD) civilians, DOD contract personnel, DOD component assets, and National Guard (NG) forces (when the Secretary of Defense [SecDef], in coordination with the governors of the affected states, elects and requests to use those forces in Title 32, United States Code, status or when federalized) in response to requests for assistance from civil authorities for domestic emergencies, law enforcement support, and other domestic activities, or from qualifying entities for special events.

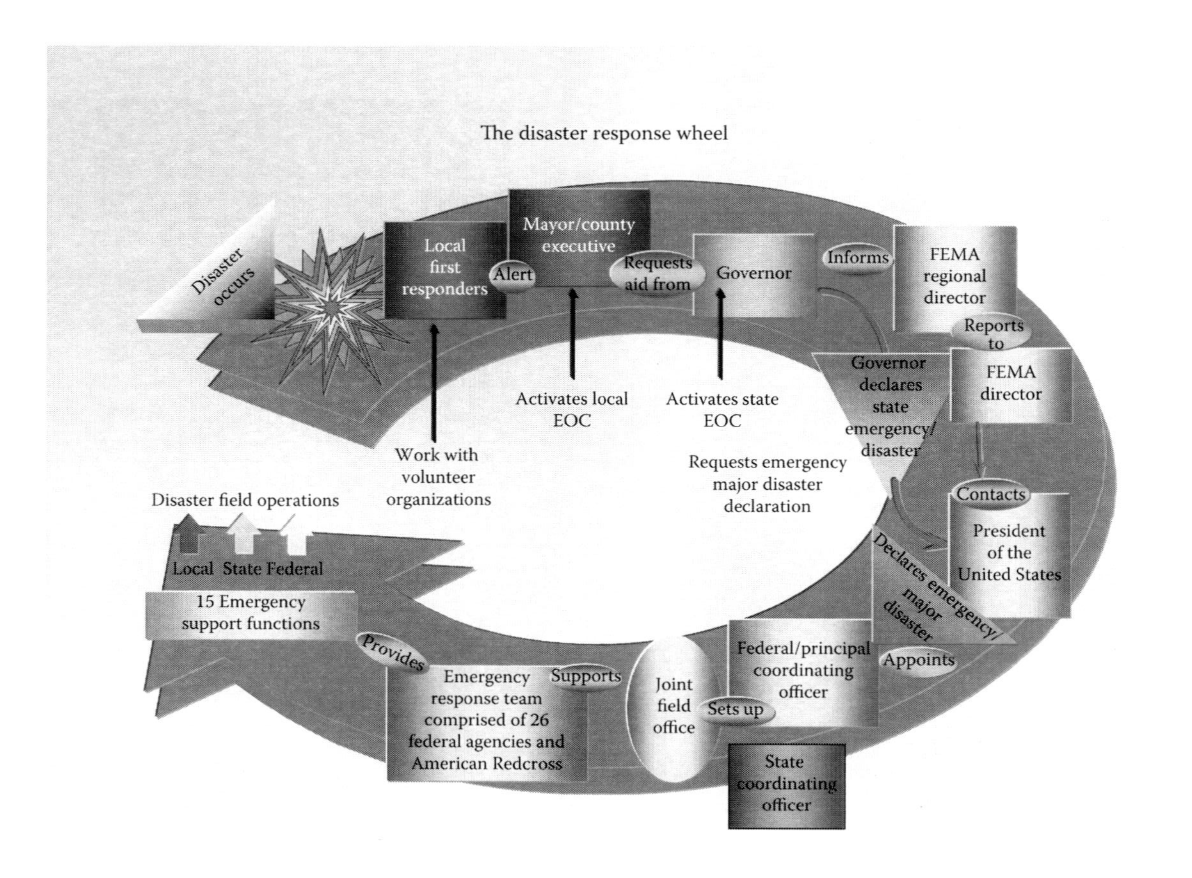

Figure 4.1 The DSCA response cycle.

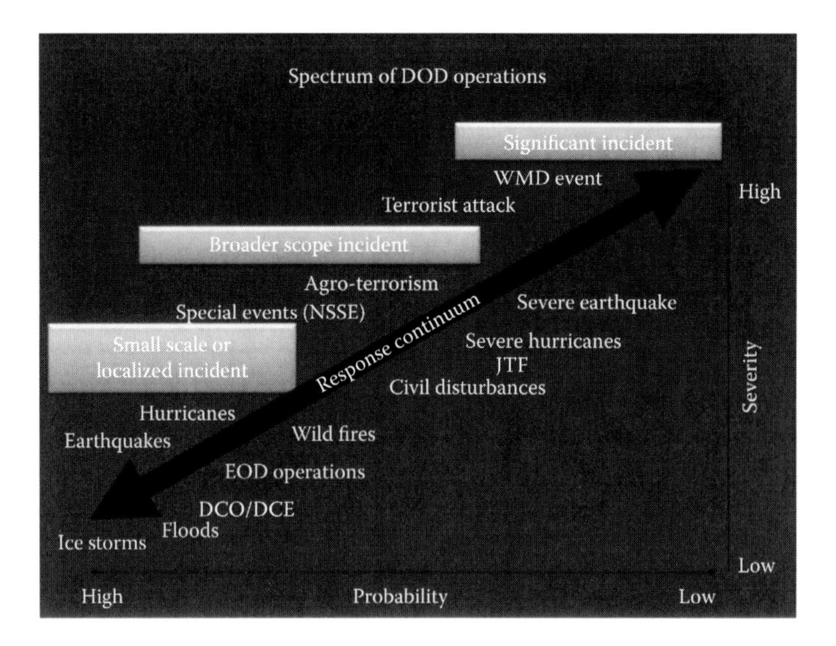

Figure 4.2 Spectrum of DOD operations.

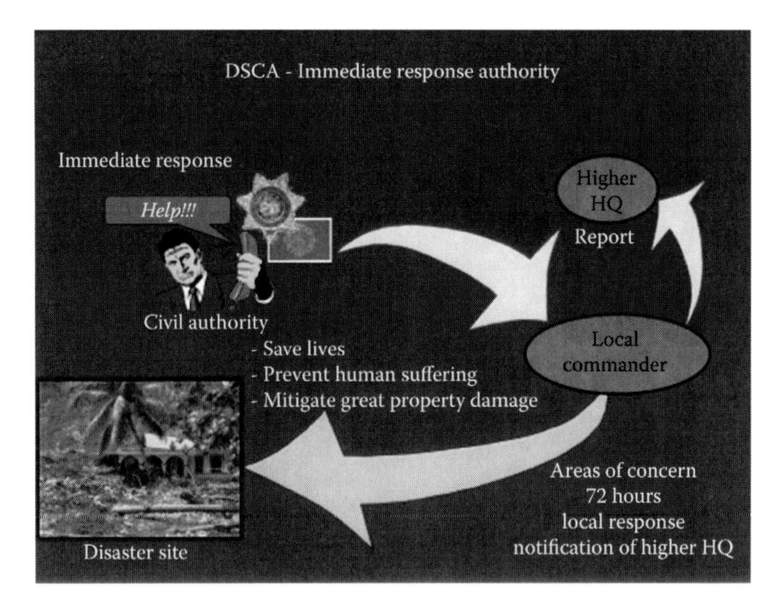

Figure 4.3 DSCA: Immediate response authority.

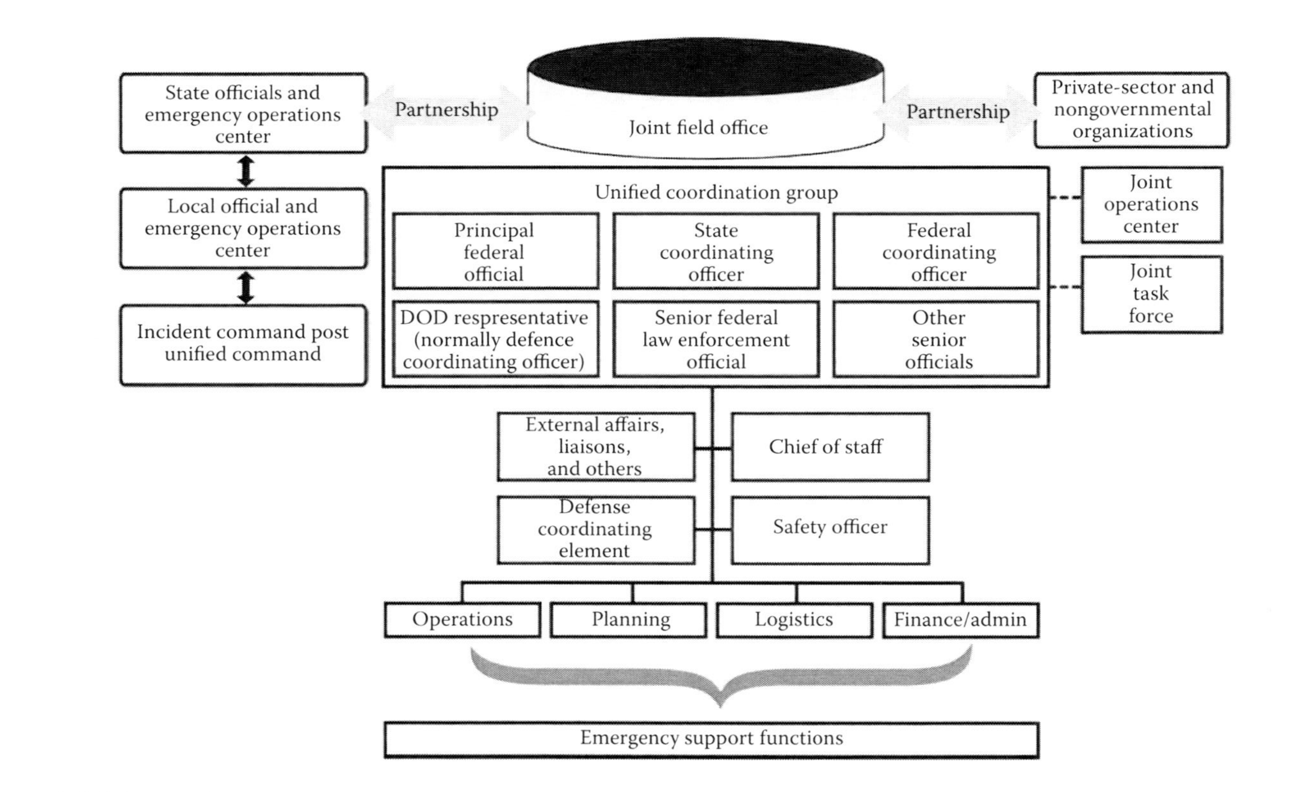

Figure 4.4 JFO organization.

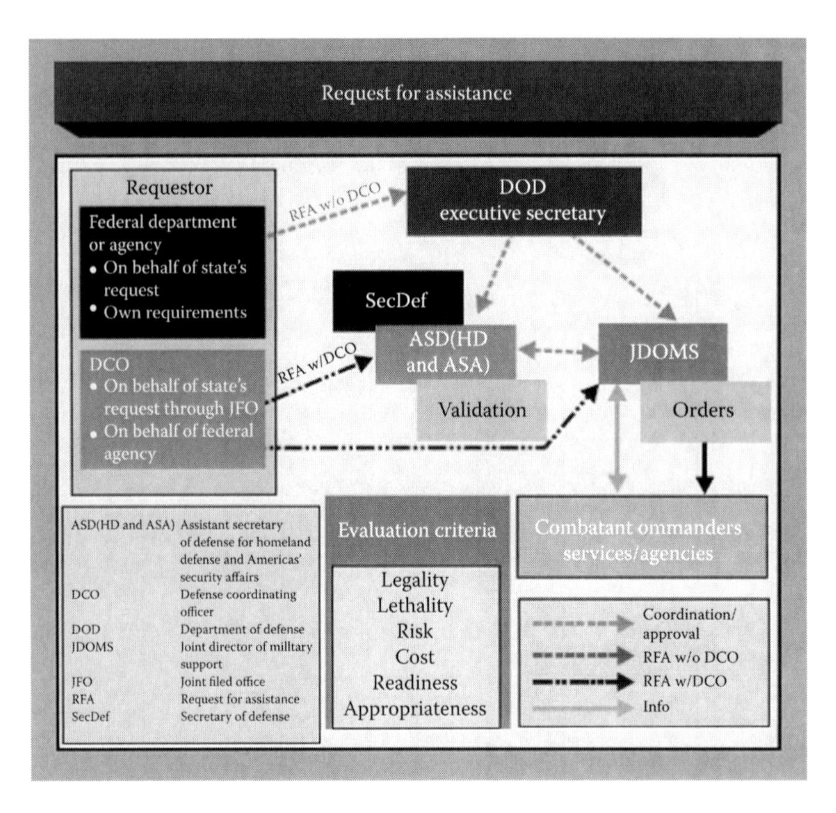

Figure 4.5 DSCA: Request for assistance.

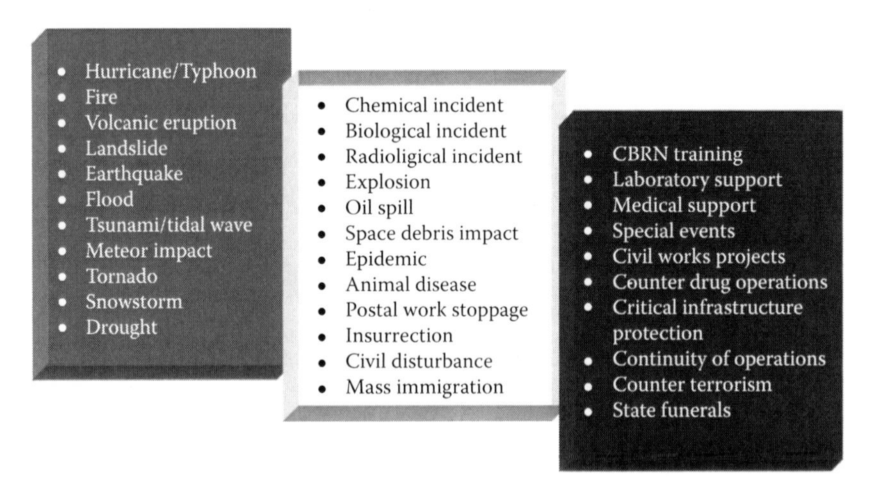

Figure 4.6 DSCA missions. The DSCA mission is to support the civilian emergency response by following the process shown in the information and chart provided.

DSCA in the United States presents a unique challenge based on the history of the country and the interaction of the federal, state, local, territorial, and tribal governments and private and non-profit organizations. These relationships establish the multiple layers and mutually reinforcing structures throughout the state and territorial governments for interaction based on the U.S. Constitution, as well as on common law and traditional relationships.

chapter five

Introduction to weapons of mass destruction (WMDs)

A "weapon of mass destruction" (WMD) is a nuclear, radiological, biological, chemical, or other weapon that can kill and bring significant harm to a large number of humans, or cause great damage to man-made structures (e.g., buildings), natural structures (e.g., mountains), or the biosphere. The scope and application of the term has evolved and been disputed, often signifying more political than technical differences. Coined in reference to aerial bombing with chemical explosives, it has come to also describe large-scale weaponry of other technologies, including biological, radiological, and nuclear. Terrorism involving weapons of mass destruction is an ever-present threat in today's world

This chapter discusses the threats posed by chemical, biological, radiological, nuclear and explosive (CBRNE) agents, toxic industrial chemicals, and toxic industrial materials.

Weapons of mass destruction

- Are chiefly designed to incite terror, not to kill
- Consist of a variety of different agents
- Can be delivered through a variety of different means
- Can be extremely difficult to control
- Are designed to cause *widespread* and *indiscriminate* death and destruction

Acronyms used to describe weapons of mass destruction

B NICE	*CBRNE*
B Biological	*C Chemical*
N Nuclear	*B Biological*
I Incendiary	*R Radiological*
C Chemical	*N* Nuclear
E Explosive	*E Explosive*

Chemical weapons of mass destruction

Why chemical weapons (CW) are attractive to terrorists

- They are inexpensive to manufacture and to obtain
- Simple technology is needed to produce them
- They are difficult to detect
- They are highly efficient (small quantities are needed)

As an example of nerve agent lethality, an amount of VX equal in size to one column of the Lincoln Memorial on the back of a penny would be lethal to you.

Sources of CW agents

- Foreign governments
- Internet recipes
- "Black Market" of the former Soviet Union
- U.S. chemical plants (chlorine, phosgene, etc.)
- U.S. Military stockpile
- 30,600 tons of nerve agents and vesicants at eight sites across the U.S.
- 1985 law directed DoD destroy stockpile by 2004
- Outdated and recovered CW are buried at 215 sites across the U.S.

Risks from chemical agents

- Detonation of CWA-containing munitions
- Atmospheric dispersal
- Contamination of food or water supplies
- Product tampering

Recent chemical terrorism events

- 1995: Aum Shinrikyo cult releases sarin vapor into Tokyo subway
    - 12 deaths and 5500 casualties
    - *4000 w/o clinical manifestation of injury*
- 1993: World Trade Center Bombing
    - explosive contained sufficient cyanide to contaminate entire building
    - cyanide destroyed in blast

Classification of chemical weapons

Chemical agents are classified by the toxic effects they have on the body.

Chief categories of agents

- Nerve agents
- Vesicants or blistering agents
- Choking or pulmonary agents
- Blood agents
- Incapacitating or riot-control agents

Nerve agents

Action: Irreversibly bind to acetyl cholinesterase (AChE), the enzyme that terminates the action of the neurotransmitter acetylcholine (ACh).

Leads to accumulation of acetylcholine, resulting in:

- *Muscarinic effects*: small pupils, dim vision, smooth muscle contraction, copious hyper-secretion (sweat, tears, runny nose)
- *Nicotinic effects*: skeletal muscle weakness, paralysis
- *CNS effects*: changes in mood, decreased mental status, seizures, coma, respiratory failure, and terminal arrhythmia
- Examples: sarin (GB), soman (GD), tabun (GA), VX gas

Muscarinic effects of nerve agents—S.L.U.D.G.E.

- *S*alivation
- *L*acrimation
- *U*rination
- *D*iaphoresis
- *G*I distress (diarrhea, vomiting)
- *E*mesis

Nerve agent antidote: MARK I kit

- Self-injectable needle
- Pralidoxime chloride (600 mg)
- Atropine (2 mg)

Vesicants/blister agents

- Produce severe blisters and chemical burns, affecting epithelium of the skin and respiratory tract
- Slow acting: cause death in 48–72 h
- Fatality due to:
  - impaired gas exchange (hypoxia)
  - loss of body fluids
  - secondary infection

- Skin and eyes affected first, then lungs and bone marrow
- Once symptoms have begun, decontamination is no longer effective
- Examples: mustard gas, lewisite

Pulmonary damaging agents

- Immediately irritating to the bronchial tree
- Early effects:
  - rhinitis/pharyngitis
  - tearing
  - eyelid spasm
  - upper respiratory tract irritation
- Later effects:
  - severe pulmonary toxicity
  - respiratory failure
- Examples: phosgene, chlorine

Blood agents or cyanides

- Combine with a cellular enzyme, inhibiting the body's ability to transport oxygen to vital organs
- Quick acting: cause death in minutes
- Relatively large dose needed to be effective
- *Initial effects*: rapid/deep breathing, anxiety, agitation, dizziness, weakness, nausea, muscle trembling
- *Later effects*: loss of consciousness, decreased respirations, seizures, arrhythmias
- Example: hydrogen cyanide

Riot-control agents

- Potent lacrimators and irritants
- Effects are believed to be transient, not meant to be lethal (though some deaths in asthmatics and the elderly have been documented)
- Considered more humane than the alternative
- (80 countries voted to ban RCA by the Geneva Convention)
- Examples: CN gas, CS gas, DM gas

Case study: Russia

- October 26, 2002
- 50 heavily armed Chechen insurgents hold hundreds of civilians hostage in a Moscow theater

- Russian Special Forces use fentanyl derivative to incapacitate the terrorists
- Over 100 hostages die from the gas

General treatment guidelines for all classes of chemical weapons

1. Move to fresh air
2. Supplemental oxygen
3. Remove clothing
4. Decontaminate skin
5. Restrict physical activity
6. Hospitalization/medical attention

Biological weapons of mass destruction

What is bioterrorism?

"Intentional or threatened use of viruses, bacteria, fungi or toxins from living organisms to produce death or disease in humans, animals or plants."

Why biologics are attractive to terrorists

- Some can be obtained from nature
- Potential dissemination over large geographic area
- Create panic and chaos
- Can overwhelm medical services
- Civilian populations may be highly susceptible
- High morbidity and mortality
- Difficult to diagnose and/or treat
- Some are transmitted person-to-person via aerosol

Characteristics of biological attacks

- Incident may not be recognized for weeks
- Responders and health workers are at risk of becoming casualties themselves
- Continuing effect with re-infection
- Require special training and equipment to handle
- Large numbers of "worried well" (30:1 ratio)
- Fear of the unknown

Genealogical classification of bioterrorism agents

- Bacterial agents
- Anthrax

- Brucellosis
- Cholera
- Plague, pneumonic
- Tularemia
- Q Fever

Source: USAMRIID—**U**nited **S**tates **A**rmy **M**edical **R**esearch **I**nstitute of **I**nfectious **D**iseases

CDC: Critical biological agents

Category A

- The nine highest priority agents; highest risk to national security
- Frequency is low; impact is high (speedy spread)
- Easily disseminated or spread person-to-person
- High mortality
- Greatest potential for widespread panic and social disruption

Category B

- Second highest priority agents
- Moderately easy to disseminate
- Moderate morbidity and low mortality (compared to Category A)

Category C

- Emerging pathogens that could be engineered for mass dissemination
- Readily available; easy to produce and disperse
- Potentially high morbidity and mortality

Category A bioterrorism agents

- *Variola major* (smallpox)
- *Bacillus anthracis* (anthrax)
- *Yersinia pestis* (plague)
- *Clostridium botulinum* (botulism)
- *Francisella tularensis* (tularemia)
- Ebola hemorrhagic fever
- Marburg hemorrhagic fever
- Lassa fever
- Argentine hemorrhagic fever

Category B bioterrorism agents

- *Coxiella burnettii* (Q fever)
- *Brucella* species (brucellosis)
- *Burkholderia mallei* (glanders)
- Venezuelan encephalomyelitis
- Eastern and Western equine encephalomyelitis
- Ricin toxin from *Ricinus communis* (castor beans)
- Epsilon toxin of *Clostridium perfringens*
- *Staphylococcus enterotoxin B*

Food/waterborne agents

- *Salmonella* species
- *Shigella dysenteriae*
- *Escherichia coli* O157:H7
- *Vibrio cholerae*
- *Cryptosporidium parvum*

Category C bioterrorism agents

- Nipah virus
- Hantavirus
- Tick-borne hemorrhagic fever viruses
- Tick-borne encephalitis viruses
- Yellow fever
- Multi-drug resistant tuberculosis (MDRTB)

Smallpox

History of smallpox

- Most deadly germ in all of human history
- First recorded case of bio warfare
- Last natural case in U.S.: 1947
- U.S. phased out vaccination from 1968 to 1972
- Last natural case in world: 1977
- "Eradicated" from the globe in 1980
- Two live cultures kept for research
- Only 10% of Soviet stockpile accounted for

Variola major (smallpox)

- Highly contagious virus (attack rate 90%)
- Person-to-person spread (P2P) by inhalation

- Mortality rate 35%
- Vaccine ~95% effective, can be administered up to four days after exposure
- No effective anti-viral agents

Smallpox: Clinical features

- *Prodrome*
- Acute onset fever, malaise, headache, backache, vomiting
- *Exanthem (Rash)*
- Begins on face, hands, forearms spreads to lower extremities then trunk over ~7 days
- Synchronous progression:
  - macules, vesicles, pustules, scabs
- Lesions on palms/soles

Smallpox versus chickenpox

- Incubation 7–17 days vs. 14–21 days
- Prodrome 2–4 days vs. minimal/none
- Distribution centrifugal vs. centripetal
- Scab formation 10–14 days vs. 4–7 days
- Scab separation 14–28 days vs. < 14 days

Smallpox vaccine

- Made from live *Vaccinia* virus
- Intradermal inoculation with bifurcated needle
- Scar (permanent) demonstrates successful vaccination
- Immunity *not* lifelong
- Adequate vaccine for all of U.S. population

Anthrax

Overview

- Primarily disease of herbivores
- Natural transmission to humans by contact with infected animals or contaminated animal products
- Three clinical forms
  - cutaneous (least lethal)
  - gastrointestinal
  - inhalational (most lethal)

- a.k.a. "Wool sorter's Disease"
- Soil reservoir
- Forms highly stable spores
- *No person-to-person transmission*
- Easy to manufacture, difficult to aerosolize
- History:
  - 1979—accidental release of spores from a USSR bioweapons factory, at least 66 dead
  - 2001—anthrax attacks in the United States, 11 contract inhalational anthrax, five died

Anthrax: Cutaneous

- Most common form (95%)
- Inoculation of spores *under* skin
- Small papule or ulcer surrounded by vesicles (24–28 h)
- Painless eschar with edema
- Mortality rate: 20% if untreated

Anthrax: Cutaneous Vesicle Development

Anthrax: Gastrointestinal

Anthrax: Inhalational

- Requires inhalation of 8,000–15,000 spores
- Initial symptoms "flu-like illness" (2–5 days)
- Fever, cough, myalgia, malaise
- Terminal symptoms (1–2 days)
- High fever, dyspnea, cyanosis
- Hemorrhagic mediastinitis/pleural effusion
- Rapid progression to shock/death
- Mediastinal widening on CXR
- Mortality rate: ~75% with antibiotic TX
- ~97% without antibiotic TX

Anthrax: Vaccine

- Current U.S. vaccine
- For persons 18–65 years of age
- Protective against cutaneous anthrax and possibly inhalational anthrax (animal data)
- Six-dose regimen over 18 months

- Limited availability
- Not currently administered to the civilian population

Plague

Overview

- Caused by bacterium *Yersinia pestis*
- Naturally occurring disease of rodents, rabbits, squirrels
- Three forms:
  - pneumonic (P2P spread)
  - bubonic (no P2P spread)
  - septicemic (no P2P spread)
- Famous for causing the "Bubonic Plague", a.k.a. the "Black Death"
- Infected rodent fleas bite a human victim
- Leads to characteristic swollen, tender lymph nodes
- Endemic to parts of the United States
- About 10–15 total cases/year, mainly in SW states
- Mostly bubonic (1–2 cases pneumonic)
- Difficult to acquire; difficult to weaponize
- Treated with antibiotics

Plague: Bubonic

- Inguinal, axillary, or cervical lymph nodes most commonly affected
- Sudden onset headache, malaise, myalgia, fever, tender lymph nodes
- Regional lymphadenitis (buboes)
- Possible papule, vesicle, or pustule at inoculation site
- 60% mortality if untreated

Plague: Bubonic

Plague: Pneumonic

- Person-to-person transmission by respiratory droplet
- Sudden onset headache, malaise, fever, cough
- Pneumonia progresses rapidly to dyspnea, cyanosis, hemoptysis
- Death from respiratory collapse/sepsis
- 100% mortality if untreated

Plague: Septicemic

- Secondary from bubonic or pneumonic forms
- Bacteria multiply in the blood

- Septic shock develops
- 100% mortality if untreated

Radioactive and nuclear weapons of mass destruction

Radiation versus radioactive material

- *Radiation*: energy transported in the form of particles or waves (alpha, beta, gamma)
- *Radioactive material*: material that contains atoms that spontaneously emit radiation
- Light, radio waves, and microwaves are types of radiation *(Ionizing radiation is what we are concerned about)*
- Radiation comes in four forms:
  - alpha particles
  - beta particles
  - neutron particles
  - gamma rays

Penetration abilities of different types of radiation

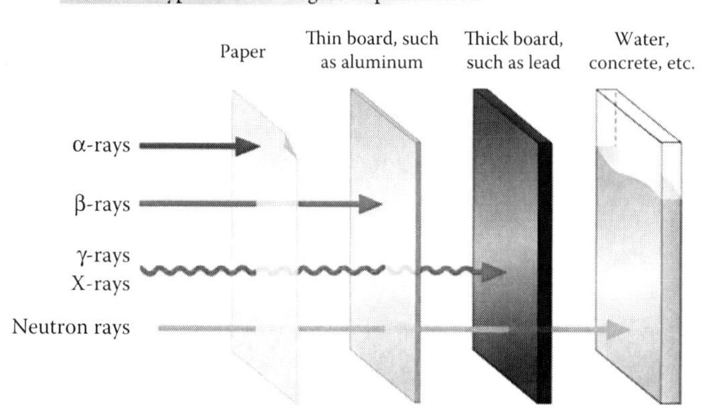

Exposure versus contamination

- *Exposure*: Irradiation of the body (gamma, X-rays, neutrons)
- *Contamination*: radioactive material on patient (external) or within patient (internal)
  - alpha
  - beta

Injuries associated with radiation exposure

- Acute radiation syndrome (ARS)
- Cutaneous radiation syndrome
- Chronic radiation exposure
- Teratogenic effects

Acute radiation syndrome

- Also known as radiation toxicity or sickness
- Requirements:
  - large, acute dose
  - penetrating
  - majority of the body is exposed
- Three classic ARS syndromes:
  - bone marrow syndrome
  - gastrointestinal syndrome
  - cardiovascular/central nervous system syndrome

Acute radiation syndrome

Cutaneous radiation syndrome

- Acute radiation exposure of the skin
- Signs/symptoms:
  - itching
  - tingling
  - erythema
  - edema
  - epilation
- Lesions may be life threatening
- Lesions do not appear for days to weeks
- Surgical treatments must be performed within 48 h to be effective

Detecting radiation: Survey meters

- Ionizing radiation interacts with detector material
- Produces a meter reading and/or audible clicks
- Typically read in counts per minute (CPM)
- Not designed for measuring radiation *exposure*

Detecting radiation: Dosimeter devices

- Self-reading dosimeter (SRD)
- Measures accumulated dose in real time
- Hold up to light and look through the eyepiece to read
- Should be checked frequently

Methods of protection

- Time
- Distance
- Shielding

Potassium iodide (KI) tablets

- Only helpful in certain cases
- Only protect the thyroid from radioactive iodine
- KI saturates the thyroid gland with stable iodine
- Must use prior to exposure to radioactive iodine

Radioactive/nuclear WMDs: Possible scenarios

- Nuclear power plant incident
- Nuclear weapon
- Improvised nuclear device (IND)
- "Dirty bomb"

Nuclear power plant incident

- Attack by air fairly easy for terrorist
- Would result in little, if any, release of radioactive material
- Redundant safety systems make catastrophic radiation leak highly unlikely

Nuclear weapon

- Manufacture requires extraordinary degree of scientific expertise
- Requires constant maintenance
- Unlikely that a terrorist organization has the resources to effectively accomplish a nuclear weapon attack

Hypothetical Scenario

Scientific American, Nov 2002: A simulated detonation of radioactive cesium-137 (3500 curies) dirty bomb at the lower tip of Manhattan Island.

Improvised nuclear device

- Weapons made from small devices that trigger uncontrolled nuclear reactions
- Difficult to manufacture
- Require frequent maintenance

"Dirty bomb"

- Radioactive/nuclear weapon of greatest concern
- Relatively easy to manufacture
- Consists of radioactive material coupled with a conventional explosive
- Immediate effect: blast injuries
- Long-term effect: chronic radiation exposure
- Would require massive decontamination effort (of people, buildings, environment)

Explosive/incendiary weapons of mass destruction

- Conventional weapons: bombs
- Incendiary devices: fire-causing devices

The most widely used WMDs!

- Explosives
- Ignite special fuels that burn extremely rapidly
- Cause a shock wave or a blast
- Cause injury by:
  - pressure wave that damages air containing organs in the body
  - throwing the body into the ground or other objects
  - propelling debris that strikes a patient causing a traumatic injury
  - causing building collapse
- Incendiary devices
- Designed to burn at extremely high temperatures
- Napalm, thermite, white phosphorous
- Cause thermal burns

Toxic industrial chemicals and toxic industrial materials

Understanding of industrial chemical agents and toxic industrial chemicals (TIC) used as weapons of mass destruction (WMD); describing selected chemical agents and TIC terms, definitions, and characteristics; and recognize the physiological signs and symptoms of exposure.

Describe toxic industrial chemicals used as weapons and the physiological signs/symptoms associated with them

- Describe choking agents and the physiological signs/symptoms associated with them
- Describe blood agents and the physiological signs/symptoms associated with them
- Describe blister agents and the physiological signs/symptoms associated with them
- Describe nerve agents and the physiological signs/symptoms associated with them

RAIN

- **R**ecognize characteristics of chemical agents
- **A**void, by protection, the hazards of chemical agents
- **I**solate the hazards of chemical agents
- **N**otify the appropriate resources and authorities when responding to a WMD event possibly involving chemical agents

Recognition

- Types
- Dissemination
- Availability
- Volatility
- Vapor density
- Odor
- Routes of entry
- General signs and/or symptoms

Toxic industrial chemicals/materials (TIC/TIM)

Using the Emergency Response Guidebook (ERG) 2012

1. *ERG:* Numbered placard for anhydrous ammonia listed by three-digit guide number
2. *ERG:* Anhydrous ammonia information listed in numerical order of ID number
3. *ERG:* Anhydrous ammonia information listed in alphabetical order by material name
4. *ERG:* Safety and emergency response recommendations according to guide number
5. *ERG:* Initial isolation and protective action distances according to material

The ERG is the upfront quick-reference guide for first responders to use in the first 30 min of any incident that involves industrial chemicals, WMDs, and radiological agents.

Though much of the press coverage focuses on the super-toxic warfare chemicals, it is very likely that a terrorist would choose a more readily available source for a WMD—toxic industrial chemicals—to cause significant casualties. One potential example is provided below, using a TIC that is readily available to communities; other TICs may also be used. Detailed information on this and other potential TICs can be found in the *ERG* and other resource materials.

Anhydrous (non-household)
Guide number 125; ID number 1005

- Dissemination: liquid or gas
- Availability: commercially available
  - used for household cleaning
  - plant growth
  - making of fertilizer
  - metal treatment operations
- Volatility: non-persistent
  - rapidly disperses
  - only poses immediate short-duration hazards
- Vapor density: lighter than air, depending upon dispersal method.

Gas particles tend to travel at high speeds in random directions

- Odor very sharp; irritating, pungent odor, similar to cat urine
- Routes of entry:
  - inhalation, ingestion, or absorption

General signs and symptoms: severe burns, coughing, nose and throat irritation, blindness, lung damage, and death. Onset of signs/symptoms usually rapid.

Sources of toxic industrial chemicals/toxic industrial materials

Toxic industrial chemicals and toxic industrial materials are chemical agents that, under certain circumstances of exposure, can cause harmful effects to living organisms.

TIC/TIM sources include:

- chemical manufacturing plants
- food processing, storage facilities with large anhydrous ammonia tanks, and chemical transportation assets

- gasoline and jet fuel storage tanks at distribution centers, airports, and barge terminals with compressed gases in tanks, pipelines, and pumping stations
- industries in which cyanide and mercury compounds are used
- pesticide manufacturing and supply distributors
- educational, medical, and research laboratories

Chemical warfare agents (using industrial chemicals)

Choking agents

Classified according to their physiological effects or their military use. In the case of choking agents, the classification is based on the physiological affect.

The following two industrial chemicals have been used as military agents, are commercially available, and could be obtained and used by terrorists:

- Phosgene (CG)
- Chlorine (Cl)

Disseminated as solid, liquid, or gas
Commercially available in

- Disinfectants
- Plastics
- Pesticides
- Solvents, chemical
- Synthesis
- Plastics
- Dyes
- Herbicides
- Volatility:
  - non-persistent
  - rapidly disperse
  - pose immediate short-duration hazards
- Vapor density heavier than air; will settle into low places
- Specific odor (newly mown hay or chlorine odor)
- Routes of entry: inhalation
- Primarily attack the airway and lungs, causing irritation
  - Fluid fills the lungs and pulmonary edema occurs
  - Known as dry-land drowning
- Onset of symptoms usually immediate

Blood agents (Cyanides)

- Hydrogen cyanide (AC)
- Cyanogen chloride (CK)

    - disseminated as liquid or gas
    - commercially available
- Used in various manufacturing processes:
    - electroplating
    - metallurgy
    - metal cleaning
    - photography
- Volatility: non-persistent,
    - rapidly disperses
    - only poses immediate short-duration hazards
    - Vapor density: ranges from slightly lighter than air to significantly heavier than air
- Odor of bitter almonds (peach pits)
- Route of entry: inhalation
- Symptoms:
    - gasping for air
    - frothing or vomiting
    - loss of consciousness, death
    - onset of symptoms very rapid, within seconds

Blister agents

- Mustards (H)
- Lewisite (L)
- Phosgene oxime (CX)

(H) are the most likely to be used, as they are the easiest to produce

- Liquid dissemination
- *Availability: not commercially available*

Nerve agents

- Tabun (GA)
- Sarin(GB)
- Soman (GD)
- VX
- Dissemination: liquid or gas
- Not commercially available

Lethality relative to chlorine

The following information is a representation of the approximate lethality of the agents in relation to chlorine. If chlorine is used as a baseline, then

- Cyanogen chloride is twice as toxic
- Phosgene is six times more toxic
- Hydrogen cyanide is seven times more toxic
- Mustard is 13 times more toxic
- Saran is 200 times more toxic
- VX is 600 times more toxic to skin; 1–2 g of mustard or sarin, or 10 mg of VX are required. VX produces skin toxicity with quantities that are 100–200 times less than either mustard or sarin.

Individuals should use the principles of time, distance, and shielding to avoid chemical agents.
First, avoid obvious hazards.
Time: minimize the time spent in the affected area.

- Evacuate the immediate area as soon as the presence of a chemical hazard is detected. Get out and stay out until the all-clear signal is given.

Distance: maximize the distance from the contaminated materials.

- The further one is from the hazard, the less likely the hazard will have any effect on life and mission capability.
- Evacuate upwind, uphill, and upstream to a distance specified in the *ERG* for the chemical agent present.

Shielding: reduce/eliminate exposure by placing an appropriate shield between the contaminant source and the individual.

- Sheltering in place
- Evacuation

Isolate

To prevent the spread of chemical agents, individuals should use standard control zones during a WMD incident.

Control Zones

The designation of safety areas at a hazardous materials incident is based the degree of hazard. Many terms are used to describe the zones utilized in a hazardous materials incident.

- *Hot Zone*—the area immediately surrounding a hazardous materials incident, which extends far enough to prevent adverse effects to personnel outside the zone; also referred to as the exclusion or restricted zone

- *Warm Zone*—the area where personnel, equipment decontamination, and hot zone support are located. Includes control points for the access corridor, and thus assists in reducing the spread of contamination. Also referred to as the decontamination, contamination reduction, or limited access corridor.
- *Cold Zone*—includes command post and other support functions deemed necessary to control the incident.

One must follow local protocols for notifying emergency services and emergency support personnel.

- What happened
- Where it happened
- When it happened
- Special hazards associated with the event
- Indicators of the type of hazard, the number of victims, and any witnesses
- Any protective measures taken
- Facilities and locations affected

Advantages/disadvantages of using chemical agents as WMDs

Ease of manufacture or acquisition and the low cost of development makes the manufacture of a chemical agent relatively simple to achieve. The ease of production of chemical agents has been compared to that of narcotics or heroin.

Chemical weapons have commonly been referred to as "the poor man's atomic bomb," due to their relative low cost and ease of manufacture.

Advantages

- Chemical agents have an immediate effect. The lethality increases as one nears the point of origin. This means the closer one gets to the source, the more incapacitated one becomes.
- Chemical agents can be odorless and/or without taste, making them harder to detect
- Chemical agents may be dispersed through explosives, sprayer systems, and in food and water contamination. They can easily spread on prevailing winds to be carried over a populated area.
- Chemical weapons tie up resources and information outlets, as they require the combined efforts of all disciplines, and increased media attention. Health care facilities, law enforcement, fire services, etc., will be exhausted during a chemical weapons attack.

- The psychological impact of chemical agents will extend far beyond the attack. Many people do not understand chemical weapons, and fear them.
- The use of chemical agents, even if supposedly non-lethal, carries the inherent danger of escalation to all-out chemical war and heightened violence.

Disadvantages

- In order to spread chemical agents effectively, large quantities are required. These may be impossible to acquire due to the Chemical Weapons Convention.
- Production and deployment can be hazardous to terrorists, who take great risks when developing weaponized chemical agents. Without proper protection, terrorists risk becoming casualties themselves.
- It is easier for America to prepare for a chemical attack as opposed to radiological, biological, or explosive attacks.

Cyberwarfare

Espionage and national security breaches

Cyber espionage is the act or practice of obtaining secrets (sensitive, proprietary, or classified information) from individuals, competitors, rivals, groups, governments, and enemies. It may be conducted for military, political, or economic advantage using illegal exploitation methods on the internet, networks, software, or computers.

Classified information that is not handled securely can be intercepted and even modified, making espionage possible from the other side of the world. Specific attacks on the United States have been given codenames like Titan Rain and Moonlight Maze.

General Alexander notes that the recently established Cyber Command is currently trying to determine whether activities such as commercial espionage or theft of intellectual property are criminal activities or actual "breaches of national security."

Sabotage

Computers and satellites that coordinate other activities are vulnerable components of a system and, if compromised, could lead to the malfunction of equipment. Compromise of military systems, such as C4ISTAR components that are responsible for orders and communications, could lead to their interception or malicious replacement.

Power, water, fuel, communications, and transportation infrastructure may all be vulnerable to disruption. The security breaches have already gone beyond stolen credit card numbers, civilians are also at risk, and potential targets can include the electric power grid, trains, or the stock market.

In mid-July 2010, security experts discovered a malicious software program called Stuxnet that had infiltrated factory computers and had spread to plants around the world. *The New York Times* considered it "the first attack on critical industrial infrastructure that sits at the foundation of modern economies."

Denial-of-service attack

In computing, a denial-of-service (DoS) attack or distributed denial-of-service (DDoS) attack is an attempt to make a machine or network resource unavailable to its intended users. Perpetrators of DoS attacks typically target sites or services hosted on high-profile web servers such as banks, credit card payment gateways, and even root name servers.

DoS attacks may not be limited to computer-based methods, as strategic physical attacks against infrastructure can be just as devastating. For example, cutting undersea communication cables may severely cripple some regions and countries with regard to their information warfare ability.

Electrical power grid

The federal government of the United States admits that the electric power grid is susceptible to cyberwarfare. The United States Department of Homeland Security works with industry to identify vulnerabilities and to help enhance the security of control system networks, while the federal government is also working to ensure that security is built-in to the next generation of "smart grid" networks being developed.

In April 2009, reports surfaced that China and Russia had infiltrated the U.S. electrical grid and had introduced software programs that could be used to disrupt the system, according to current and former national security officials. The North American Electric Reliability Corporation (NERC) has issued a public notice that warns that the electrical grid is not adequately protected from cyber-attack. China denies intruding into the U.S. electrical grid. One countermeasure would be to disconnect the power grid and to run the grid with droop speed control only. Massive power outages caused by a cyber-attack could disrupt the economy, distract from a simultaneous military attack, or create a national trauma.

Howard Schmidt, former Cyber-Security Coordinator of the United States, believes that while it is possible that hackers have gotten into administrative computer systems of utility companies, these are not linked to the equipment controlling the grid, at least not in developed countries. Schmidt has never heard that the grid itself has been hacked.

Military

General Keith B. Alexander, first head of the recently formed USCYBERCOM, told the Senate Armed Services Committee that computer network warfare is evolving so rapidly that there is a "mismatch between our technical capabilities to conduct operations and the governing laws and policies. Cyber Command is the newest global combatant and its sole mission is cyberspace, outside the traditional battlefields of land, sea, air and space." Cyber Command will attempt to find and, when necessary, neutralize, cyber-attacks and defend military computer networks.

Alexander sketched out the broad battlefield envisioned for the computer warfare command, listing the kind of targets that his new headquarters could be ordered to attack, including "traditional battlefield prizes—command-and-control systems at military headquarters, air defense networks, and weapons systems that require computers to operate."

One cyberwarfare scenario, Cyber Shockwave, which was wargamed at the cabinet level by former administration officials, raised issues ranging from the National Guard, to the power grid, to the limits of statutory authority. The distributed nature of internet-based attacks means that it is difficult to determine motivation and attacking party, and that it is unclear when a specific act should be considered an act of war.

Examples of cyberwarfare driven by political motivations can be found worldwide. In 2008, Russia began a cyber-attack on the Georgian government website, which was carried out along with Georgian military operations in South Ossetia. In 2008, Chinese "nationalist hackers" attacked CNN as it reported on Chinese repression on Tibet.

Terrorism

Eugene Kaspersky, founder of Kaspersky Lab, concludes that "cyber terrorism" is a more accurate term than "cyber war." He states that "with today's attacks, you are clueless about who did it or when they will strike again. It's not cyber-war, but cyber terrorism." He also equates large-scale cyber weapons, such as the Flame Virus and Net Traveler Virus, which his company discovered, to biological weapons, claiming that in an interconnected world, they have the potential to be equally destructive.

Civil

Potential targets of internet sabotage include all aspects of the internet, from the backbones of the web, to internet service providers, to the varying types of data-communication mediums and network equipment. This would include web servers, enterprise information systems, client/server systems, communication links, network equipment, and the desktops and laptops in businesses and homes. Electrical grids and telecommunication systems are also deemed vulnerable, especially due to current trends in automation.

Private sector

Computer hacking represents a modern threat in ongoing industrial espionage, and as such, is presumed to occur widely. Typically, this type of crime is underreported. According to McAfee's George Kurtz, corporations around the world face millions of cyber-attacks a day. "Most of these attacks don't gain any media attention or lead to strong political statements by victims." This type of crime is usually financially motivated.

Non-profit research

However, not all cyberwarfare issues are motivated by achieving profit or personal gain. There are still institutes and companies, like the University of Cincinnati or the Kaspersky Security Lab, which are trying to increase awareness of this topic by researching and publishing information on new security threats.

National Incident Management System/ Incident Command System (NIMS/ICS)

The National Incident Management System (NIMS) identifies concepts and principles that answer how to manage emergencies, from preparedness to recovery, regardless of their cause, size, location, or complexity. NIMS provides a consistent, nationwide approach and a vocabulary for multiple agencies or jurisdictions to work together to build, sustain, and deliver the core capabilities needed to achieve a secure and resilient nation.

Consistent implementation of NIMS provides a solid foundation across jurisdictions and disciplines to ensure effective and integrated preparedness, planning, and response. NIMS empowers the components of the National Preparedness System, a requirement of Presidential Policy Directive (PPD)-8, to guide activities within the public and private sector, and describes the planning, organizing, equipping, training, and exercising needed to build and sustain the core capabilities in support of the *National Preparedness Goal*.

The *National Preparedness Goal* is to become a secure and resilient nation with the capabilities required across the whole community to prevent, protect against, mitigate, respond to, and recover from the threats and hazards that pose the greatest risk. To achieve the Goal, existing preparedness networks and activities, such as NIMS, will be used to improve training and exercise programs, promote innovation, and ensure that the administrative, financial, and logistical systems are in place to support these capabilities.

National Integration Center (NIC) guidance delivery and whole-community engagement

The National Integration Center (NIC) mission requires the development of preparedness-related doctrine, policy, and guidance to reflect the collective expertise and experience of the whole community. Additionally, the NIC has a duty to ensure its products support "on-the-ground" stakeholders that will be responsible for changing outcomes for survivors and their communities. In order to achieve a national scope and perspective, it is critical that stakeholders be the source of content for national doctrine and guidance to the greatest extent possible.

To that end, through separate and distinct national engagement and public comment processes, the NIC

- Identifies and validates needs and requirements to deliver the core capabilities across the prevention, protection, mitigation, response, and recovery mission area communities of practice at the local, state, territorial, tribal, and federal levels.
- Works to attain a national perspective in support of the National Preparedness System, to achieve the National Preparedness Goal.
- Seeks to engage the nation early and often, through a whole-community approach that increases ownership and subsequent adoption of national doctrine, policy, and guidance.

Comments received from national engagement are adjudicated by the NIC through collaborative engagement with practitioners and subject-matter experts. This all-of-nation approach results in whole-community-informed documents that can be used in the public comment period—a second and more formal "national engagement" that results in additional whole-community input.

National engagement and public comment periods provide the opportunity to facilitate the national scope, perspective, and operational viability of national doctrine, guidance, and policy.

National Integration Center (NIC)

The National Integration Center (NIC) has primary responsibility for the maintenance and management of national preparedness doctrine, to include the Goal, NPS, NIMS, and the National Planning Frameworks. The NIC provides a national perspective and strategic direction for the NPS and NIMS, as well as the routine maintenance and continuous improvement of these systems. The NIC develops strategies, doctrine, policies, guidance, and best practices in collaboration with practitioners and subject-matter experts from the whole community, across the five mission areas.

Additionally, the NIC facilitates, develops, and issues national standardized resource-typing definitions for resources commonly requested during situations necessitating interstate mutual aid. The NIC uses both measurable standards (including accepted industry and discipline standards) and input from stakeholders to ensure that these typed resources accurately reflect operational capabilities. Three key documents that speak to the NIC's authority to facilitate, coordinate, and/or conduct NIMS resource typing are the following:

- *Post-Katrina Emergency Management Reform Act of 2006*
- *National Incident Management System (December 2008)*
- *Presidential Policy Directive 8: National Preparedness (March 2011)*

The NIC relies on its Strategic Resource Group—practitioners and subject-matter experts from state, tribal, and local governments; nongovernmental organizations (NGOs); and the private sector—to assist with resource-typing definitions, including job title/position qualifications for the credentialing of personnel. Local, state, territory, tribal and federal agencies and departments depend upon these definitions to identify and inventory their equipment, personnel, and teams. NIMS resource-management guidance is continually adjusted to reflect currency gained through lessons learned and after-actions from incidents and events.

Incident Command System

The Incident Command System (ICS) is a standardized, on-scene, all-hazards incident-management approach that

- Allows for the integration of facilities, equipment, personnel, procedures, and communications operating within a common organizational structure.
- Enables a coordinated response among various jurisdictions and functional agencies, both public and private.
- Establishes common processes for planning and managing resources.

ICS is flexible, and can be used for incidents of any type, scope, and complexity. ICS allows its users to adopt an integrated organizational structure to match the complexities and demands of single or multiple incidents.

ICS is used by all levels of government—federal, state, tribal, and local—as well as by many nongovernmental organizations and the private sector.

ICS is also applicable across disciplines. It is typically structured to facilitate activities in five major functional areas: command, operations, planning, logistics and finance/administration. All of the functional areas may or may not be used, based on the incident needs. Intelligence/investigations is an optional, sixth functional area that is activated on a case-by-case basis.

As a system, ICS is extremely useful; not only does it provide an organizational structure for incident management but it also guides the process for planning, building and adapting that structure. Using ICS for every incident or planned event helps hone and maintain skills needed for large-scale incidents.

ICS organization

The Incident Command System comprises five major functional areas: *ICFLOP*.

- Command
- Operations
- Planning
- Logistics
- Finance/Administration

(A sixth functional area, intelligence/investigations, may be established if required.)

Modular expansion

The ICS organizational structure is modular, extending to incorporate all elements necessary for the type, size, scope, and complexity of an incident. It builds from the top down; responsibility and performance begin with Incident Command. When the need arises, four separate sections can be used to organize the General Staff. Each of these sections may have several subordinate units, or branches, depending on the incident's management requirements. If one individual can simultaneously manage all major functional areas, no further organization is required. If one or more of the functions requires independent management, an individual is assigned responsibility for that function.

To maintain a manageable span of control, the initial responding Incident Commander (IC) may determine it necessary to delegate functional management to one or more Section Chiefs. The Section Chiefs may further delegate management authority for their areas, as required. A Section Chief may establish branches, groups, divisions, or units, depending on the section. Similarly, each functional Unit Leader will further assign individual tasks within the unit, as needed.

The use of deputies and assistants is a vital part of both the organizational structure and the modular concept. The IC may have one or more deputies, who may be from the same or an assisting agency. Deputies may also be used at section and branch levels of the organization. a deputy, whether at the command, section, or branch level, must be fully qualified to assume the position.

Deputy ICs are primarily used as subordinates to the Command Staff, which includes the Public Information Officer, Safety Officer, and Liaison Officer.

They have levels of technical capability, qualifications, and responsibility subordinate to the primary positions. The modular concept described above is based on the following considerations:

- To perform specific tasks as requested by the IC.
- To perform the incident command function in a relief capacity (e.g., to take over the next operational period; in this case, the deputy will then assume the primary role).
- To represent an assisting agency that may share jurisdiction or have jurisdiction in the future.
- Developing the organization's structure to match the function or task to be performed.
- Staffing only the functional elements required to perform the task.
- Implementing recommended span-of-control guidelines (Table 5.1).

Table 5.1 ICS command chart

Organizational elements	Leadership position title	Support positions
Incident command	Incident commander	Deputy
Command staff	Officer	Assistant
Section	Section chief	Deputy
Branch	Branch director	Deputy
Divisions and groups	Supervisors	N/A
Unit	Unit leader	Manager, coordinator
Strike team/task force	Leader	Single resource boss companies/crews
Technical specialist	Specialist	N/A
Single resource boss	Boss	N/A

Command staff

In an ICS organization, Incident Command consists of the *Incident Commander* and various Command Staff positions.

The *Public Information Officer* is responsible for interfacing with the public and media and with other agencies with incident-related information requirements.

The *Safety Officer* monitors incident operations and advises Incident Command on all matters relating to operational safety, including the health and safety of emergency responder personnel.

The *Liaison Officer* is Incident Command's point of contact for representatives of other governmental departments and agencies, NGOs, and/or the private sector (with no jurisdiction or legal authority) to provide input on their organization's policies, resource availability, and other incident-related matters.

Command functions/sections

The *Operations Section* is responsible for managing operations directed toward reducing the immediate hazard at the incident site, saving lives and property, establishing situation control, and restoring normal conditions.

Incidents can include acts of terrorism, wild land and urban fires, floods, hazardous material spills, nuclear accidents, aircraft accidents, earthquakes, hurricanes, tornadoes, tropical storms, war-related disasters, public health and medical emergencies, and other incidents requiring an emergency response.

The *Planning Section* is responsible for collecting, evaluating, and disseminating operational information pertaining to the incident. This section maintains information and intelligence on the current and forecasted situation, as well as the status of resources assigned to the incident. The Planning Section prepares and documents Incident Action Plans and incident maps, and gathers and disseminates information and intelligence critical to the incident. The Planning Section has four primary units, and may also include technical specialists to assist in evaluating the situation and forecasting requirements for additional personnel and equipment.

The *Logistics Section* provides for all the support needs for the incident, such as ordering resources and providing facilities, transportation, supplies, equipment maintenance and fuel, food service, communications, and medical services for incident personnel. The Logistics Section is led by a Section Chief, who may also have one or more deputies. Having a deputy is encouraged when all designated units are established at an incident site. When the incident is very large or requires a number of facilities with large numbers of equipment, the Logistics Section can be divided into branches. This helps with span of control by providing more effective supervision and coordination among the individual units. Conversely, in smaller incidents or when fewer resources are needed, a branch configuration may be used to combine the task assignments of individual units.

The *Finance/Administration Section* is established when there is a specific need for financial and/or administrative services to support incident management activities. Large or evolving scenarios involve significant funding originating from multiple sources. In addition to monitoring multiple sources of funds, the Section Chief must track and report to the IC/UC the accrued cost as the incident progresses. This allows the IC/UC to forecast the need for additional funds before operations are affected negatively, and it is particularly important if significant operational resources are under contract from the private sector.

Establishing branches to maintain recommended span of control

The recommended span of control for the Operations Section Chief, as for all managers and supervisory personnel, is 1:5—as high as 1:10 for larger-scale law enforcement operations. When this is exceeded, the Operations Section Chief should set up two branches, allocating the divisions and groups between them.

First responder categories and capabilities

First responder awareness level guidelines

Law enforcement

Address training requirements for law enforcement personnel who are likely to witness or discover an event involving the terrorist/criminal use of weapons of mass destruction or who may be sent out to initially investigate the report of such an event. Generally, all actions to be taken by these personnel should be conducted from within the cold zone. If personnel find themselves in the warm or hot zones, they are to move from that zone and encourage others, if ambulatory, to move to a staging area away from the immediate threat. They should attempt to minimize further contamination.

Awareness level guidelines for law enforcement officers

- Recognize hazardous materials incidents.
- Know the protocols used to detect the potential presence of weapons of mass destruction (WMD) agents or materials.
- Know and follow self-protection measures for WMD events and hazardous materials events.
- Know procedures for protecting a potential crime scene.

- Know and follow agency/organization's scene security and control procedures for WMD and hazardous material events.
- Possess and know how to properly use equipment to contact dispatcher or higher authorities to report information collected at the scene and to request additional assistance or emergency response personnel.

Fire fighters

Address training requirements for fire fighters who are likely to witness or discover an event involving the terrorist/criminal use of weapons of mass destruction (WMD) or who may be sent out to initially investigate the report of such an event. Generally, all actions to be taken by these personnel should be conducted from within the cold zone. If personnel find themselves in the warm or hot zone, they are to move from that zone and encourage others, if ambulatory, to move to a staging area away from the immediate threat.

- Recognize hazardous materials incidents.
- Know the protocols used to detect the potential presence of WMD agents or materials.
- Know and follow self-protection measures for WMD events and hazardous materials events.
- Know procedures for protecting a potential crime scene.
- Know and follow agency/organization's scene security and control procedures for WMD and hazardous material events.
- Possess and know how to properly use equipment to contact dispatcher or higher authorities to report information collected at the scene and to request additional assistance or emergency response personnel.

Emergency medical service (EMS) providers

Address training requirements for EMS providers who are likely to respond to or discover an event involving the terrorist/criminal use of weapons of mass destruction or who may be sent out to initially investigate the report of such an event. Generally, all actions to be taken by these personnel should be conducted from within the cold zone. If personnel find themselves in the warm or hot zone, they are to move from that zone and encourage others, if ambulatory, to move to a staging area away from the immediate threat. They should attempt to minimize further contamination. It is assumed that the EMS provider at this awareness level does not have emergency response supplies with him/her when arriving at the potential WMD scene (unless dispatched). The EMS provider that is anticipated to be covered by these guidelines would be trained in first aid and cardiopulmonary resuscitation (CPR), using the Red Cross Community

First Aid Course or equivalent, up to and including paramedic-trained personnel and emergency physicians.

- Recognize hazardous materials incidents.
- Know the protocols used to detect the potential presence of WMD agents or materials.
- Know and follow self-protection measures for WMD events and hazardous materials events.
- Know procedures for protecting a potential crime scene.
- Know and follow agency/organization's scene security and control procedures for WMD and hazardous material events.
- Possess and know how to properly use equipment to contact dispatcher or higher authorities to report information collected at the scene and to request additional assistance or emergency response personnel. Know how to characterize a WMD event and be able to identify available response assets within the affected jurisdiction(s).

Emergency management personnel

Address training requirements for emergency management personnel who are likely to witness or discover an event involving the terrorist/criminal use of weapons of mass destruction or who may be sent out to initially investigate the report of such an event. Generally, all actions to be taken by these personnel should be conducted from within the cold zone. If personnel find themselves in the warm or hot zone, they are to move from that zone and encourage others, if ambulatory, to move to a staging area away from the immediate threat.

- Recognize hazardous materials incidents.
- Know the protocols used to detect the potential presence of WMD agents or materials.
- Know and follow self-protection measures for WMD events and hazardous materials events.
- Know procedures for protecting a potential crime scene.
- Know and follow agency/organization's scene security and control procedures for WMD and hazardous material events.
- Possess and know how to properly use equipment to contact dispatcher or higher authorities to report information collected at the scene and to request additional assistance or emergency response personnel.

Public works employees

Address training requirements for public works employees who are likely to witness or discover an event involving the terrorist/criminal use of WMD or who may be sent out to initially investigate the report of such an event. This training should target all non-operational employees; line personnel and operations supervisors, including highway maintenance crews; planners, engineers, and lab technicians; and superintendents. This training also is appropriate for agency directors employed by public works facilities associated with a local community, including a public works facility such as wastewater treatment or water operations covered by the emergency response plan. Generally, all actions to be taken by these personnel should be conducted from within the cold zone. If personnel find themselves in the warm or hot zone, they are to move from that zone and encourage others, if ambulatory, to move to a staging area away from the immediate threat.

- Recognize hazardous materials incidents.
- Know the protocols used to detect the potential presence of WMD agents or materials.
- Know and follow self-protection measures for WMD events and hazardous materials events.
- Know procedures for protecting a potential crime scene.
- Know and follow agency/organization's scene security and control procedures for WMD and hazardous material events.
- Possess and know how to properly use equipment to contact dispatcher or higher authorities to report information collected at the scene and to request additional assistance or emergency response personnel.

First responders' performance level guidelines

Law enforcement

This level is divided into two parts, with a separate set of training guidelines for each part. The training guidelines target officers who will likely be responding to the scene of a hazardous materials event or a potential terrorist criminal use of WMD. These officers will conduct on-scene operations within the warm zone and/or the hot zone (if properly trained and equipped) that has been set up on the scene of a potential WMD or hazardous materials event to control and close out the incident. It is expected that those officers trained for Performance Level A will work in the warm and cold zones and support those officers working in the hot zone. Officers trained for Performance Level B will work in the hot zone, and in the other zones set up on the incident scene as needed.

Performance Level A (operations level) guidelines for law enforcement officers

- Have successfully completed adequate and proper training at the awareness level for events involving hazardous materials and for WMD and other specialized training.

- Know the Incident Command System and be able to follow Unified Command System procedures for the integration and implementation of each system. Know how the systems integrate and support the incident. Be familiar with the overall operation of the two command systems and be able to assist in implementation of the Unified Command System if needed.
- Know and follow self-protection measures and rescue and evacuation procedures for WMD events.
- Know and follow procedures for working at the scene of a potential WMD event.

Performance Level B (technician level) guidelines for law enforcement officers

- Have successfully completed training at Awareness Level and Performance Level A for events involving hazardous materials, and for WMD and other specialized training.
- Know and follow self-protection measures and rescue and evacuation procedures for WMD events.
- Know and follow procedures for performing specialized work at the scene of a potential WMD event.
- Know and follow Incident Command System and Unified Command System procedures and steps required for implementation of each system. Understand how the two systems are to work together.

Fire service

This level is divided into two parts, with a separate set of training guidelines for each part. The training guidelines target fire fighters who will likely be responding to the scene of a hazardous materials event or a potential terrorist/criminal use of weapons of mass destruction (WMD) event. These fire fighters will conduct on-scene operations within the warm zone and/or the hot zone (if properly trained and equipped) that has been set up on the scene of a potential WMD or hazardous materials event to control and close out the incident. It is expected that fire fighters trained for Performance Level A will work in the warm and cold zones and support those fire fighters working in the hot zone. Fire fighters trained for Performance Level B will work in the hot zone, and in the other zones set up on the incident scene as needed.

Performance Level A (operations level) guidelines for fire fighters

- Have successfully completed adequate and proper training at the awareness level for events involving hazardous materials and for WMD and other specialized training.

- Know the Incident Command System and be able to follow Unified Command System procedures for the integration and implementation of each system. Know how the systems integrate and support the incident. Be familiar with the overall operation of the two command systems and be able to assist in implementation of the Unified Command System if needed.
- Know and follow self-protection measures and rescue and evacuation procedures for WMD events.
- Know and follow procedures for working at the scene of a potential WMD event.

Performance Level B (technical level) guidelines for fire fighters

- Have successfully completed training at Awareness Level and Performance Level A for events involving hazardous materials, and for WMD and other specialized training.
- Know and follow self-protection measures and rescue and evacuation procedures for WMD events.
- Know and follow procedures for performing specialized work at the scene of a potential WMD event.
- Know and follow Incident Command System and Unified Command System procedures and steps required for implementation of each system. Understand how the two systems are to work together.

EMS

This level is divided into two parts, with a separate set of training guidelines for each part. The training guidelines for EMS providers target personnel who will likely be responding to the scene of a hazardous materials event or a potential terrorist/criminal use of WMD event. These EMS responders will conduct on-scene operations within the warm and cold zones that have been set up on the scene of a potential WMD or hazardous materials event. They are expected to provide emergency medical assistance and treatment to the victims, support those involved in the control and mitigation of the on-scene hazards, and to assist in bringing the incident to a successful conclusion. EMS responders trained for Performance Level A will work in the warm and cold zones and support the other emergency responders in any of the three zones. EMS responders trained for Performance Level B will supervise or serve as team leaders for EMS groups given various assignments by the incident commander or incident management team under the Unified Command System. Performance Level B EMS responders will work in the warm and cold zones, but they will have some special training in rescuing or assisting in rescuing victims in the hot zone. Hot-zone rescue efforts will be coordinated with the fire service and hazardous material (HAZMAT) responders.

Performance Level A (operations level) guidelines for EMS responders

- Have successfully completed adequate and proper training at the awareness level for events involving hazardous materials, and for WMD and other specialized training.
- Know the Incident Command System and be able to follow Unified Command System procedures for the integration and implementation of each system. Know how the systems integrate and support the incident. Be familiar with the overall operation of the two command systems and be able to assist in implementation of the Unified Command System if needed.
- Know and follow self-protection measures and rescue and evacuation procedures for WMD events.
- Know and follow procedures for working at the scene of a potential WMD event.

Performance Level B (technical level) guidelines for fire fighters

- Have successfully completed training at Awareness Level and Performance Level A for events involving hazardous materials, and for WMD and other specialized training.
- Know and follow self-protection measures and rescue and evacuation procedures for WMD events.
- Know and follow procedures for performing specialized work at the scene of a potential WMD event.
- Know and follow Incident Command System and Unified Command System procedures and steps required for implementation of each system. Understand how the two systems are to work together.

Hazardous materials

This performance level will be a single tier of training guidelines for HAZMAT emergency responders. These guidelines target those emergency responders who will be responding to the scene of a HAZMAT or potential terrorist/criminal use of WMD event. These HAZMAT responders will conduct on-scene operations within the hot and warm zones that have been established at the scene of a potential WMD or HAZMAT event. These personnel may also work in the cold zone as needed. These HAZMAT responders will likely be involved controlling and mitigating hazards found on the scene and in bringing the incident to a successful conclusion under the direction of an operations officer and the on-scene incident commander.

Performance Level A (operations level) guidelines for HAZMAT responders

- Have successfully completed adequate and proper training at the awareness level for events involving HAZMAT, and for WMD and other specialized training.

- Know the Incident Command System and be able to follow Unified Command System procedures for the integration and implementation of each system. Know how the systems integrate and support the incident. Be familiar with the overall operation of the two command systems and be able to assist in implementation of the Unified Command System if needed.
- Know and follow self-protection measures and rescue and evacuation procedures for WMD events.
- Know and follow procedures for working at the scene of a potential WMD event.

Performance Level B (technical level) guidelines for HAZMAT responder

- Have successfully completed training at Awareness Level and Performance Level A for events involving HAZMAT and for WMD and other specialized training.
- Know and follow self-protection measures and rescue and evacuation procedures for WMD events.
- Know and follow procedures for performing specialized work at the scene of a potential WMD event.
- Know and follow Incident Command System and Unified Command System procedures and steps required for implementation of each system. Understand how the two systems are to work together.

Public works

This performance level addresses training requirements for all non-operational employees; line personnel and operations supervisors; planners, engineers, and lab technicians; and superintendent/agency directors employed by public works facilities. These personnel will be involved in a community response to a WMD incident, particularly an incident affecting wastewater treatment or water operations, which may represent WMD targets. It is assumed that non-public works personnel will comprise emergency responders. Therefore, training requirements associated with any Federal or State contingency planning and preparedness requirements for responding to such an incident are not considered.

Performance Level A (operations level) guidelines for public works personnel

- General line operations personnel and supervisors: Have successfully completed additional training beyond awareness level to be able to provide skilled support services in the event of a WMD attack targeting a public works facility.
- Planners, engineers, and lab technicians: Have successfully completed additional training to effectively respond to a WMD incident, either within a public works facility or within the community.

Planning and management level guidelines

Law enforcement

Address training requirements for law enforcement officials who are expected to be part of the leadership and management team that likely will respond to an event involving the terrorist/criminal use of WMD. At the very least, law enforcement managers will be involved in onsite planning for and managing scene security services. They will help set up the crime scene investigation and evidence gathering that will be coordinated with the command post at the scene. These personnel are expected to manage onsite law enforcement resources and assist the incident commander in bringing the event to a successful conclusion. Generally, all of the actions to be taken by these law enforcement managers should be conducted from within the cold zone. As access is provided to law enforcement officers to conduct their potential crime scene investigation, there may be times when law enforcement managers come within the warm zone. It is expected that law enforcement managers will be integrated into the overall command structure that is implemented for the management and supervision of resources and assets being deployed to mitigate and recover from the overall WMD emergency event.

Planning and management level guidelines for law enforcement managers

- Have successfully completed training in awareness, performance, and management levels for events involving hazardous materials and for WMD.
- Know Incident Command System and the Unified Command System procedures and the steps required for implementation of each system. Understand how the systems are integrated and implemented to work together and what information the on-scene manager needs from the law enforcement manager. Be familiar with the full range of incident command functions and be able to fulfill any functions related to law enforcement operations.
- Know protocols to secure and retain control of the emergency scene and to allow only authorized persons involved with the emergency incident to gain access to the scene of WMD agents and/or hazardous materials.
- Know and follow self-protection measures and protective measures for personnel on the scene of WMD events and hazardous materials events.
- Know and follow procedures for protecting a potential crime scene.
- Know plans and assets available for the crime scene investigation and control of WMD and HAZMAT events to secure and retain evidence removed from the scene.

Fire service

Address training requirements for fire department senior officers who are expected to be part of the leadership and management team that likely will respond to an event involving the terrorist/criminal use of WMD. Fire department senior officers will be involved in planning for and managing the emergency onsite scene and will help implement the on-scene command post. These officers are expected to manage fire fighters and other allied emergency responders, who will support the ongoing operations to mitigate and control the hazardous agents and materials, using any available resources to safely and sufficiently conclude the event. Generally, actions to be taken by fire department senior officers should be conducted from the cold zone (sometimes from the warm zone). It is expected that fire service managers will be integrated into the overall command structure that is implemented for the management and supervision of resources and assets being deployed to mitigate and recover from the overall WMD emergency event.

Planning and management level guidelines for fire department senior officers

- Have successfully completed training in awareness, performance, and management levels for events involving HAZMAT and for WMD.
- Know Incident Command System and the Unified Command System procedures and the steps required for implementation of each system.
- Understand how the systems are integrated and implemented to work together and what information the on-scene manager needs from the fire department manager. Be familiar with the full range of incident command functions, and be able to fulfill any functions related to fire department operations.
- Know protocols to secure, mitigate, and remove hazardous agents or materials that may be WMD agents or materials.
- Know and follow self-protection and protective measures for emergency responders to WMD and HAZMAT events.
- Understand development of the Incident Action Plan and know assets available for controlling WMD and HAZMAT events, in coordination with the on-scene incident commander. In collaboration with the on-scene incident commander, be able to assist in planning and in determining operational goals and objectives to bring the event to a successful conclusion.
- Know and follow procedures for protecting a potential crime scene.
- Know and follow department protocols for medical monitoring of response personnel involved with or working at WMD and HAZMAT events.

EMS

Address training requirements for emergency providers who will be part of the leadership and management of the emergency medical team likely to respond to an event involving the terrorist/criminal use of WMD. These emergency medical managers will be involved in planning for and managing onsite emergency medical services (EMS).

They also will help set up the command post at the scene. These personnel are expected to manage emergency medical resources used to successfully conclude the event. Generally, all of the actions to be taken by these emergency medical team managers should be conducted from within the cold zone (and at times from within the warm zone). It is expected that the emergency medical team managers will be integrated into the overall command structure set up for management and supervision of resources and assets deployed to control and conclude the overall WMD or HAZMAT emergency event.

Planning and management level guidelines for emergency medical providers team managers

- Have successfully completed training in awareness, performance, and management levels for events involving HAZMAT and for WMD.
- Know and follow Incident Command System and Unified Command System procedures and requirements for implementing each system. Understand how the systems are implemented and integrated. Know what information the on-scene incident commander will need from the EMS manager.
- Know and follow protocols to provide emergency medical treatment to persons involved in a potential or actual WMD event.
- Know and follow self-protection and protective measures for victims of WMD events and HAZMAT events. Understand the special hazards to humans from WMD agents and hazardous materials.
- Know the plans and assets available for transporting the victims of WMD and hazardous materials events to more advanced medical care at hospitals and similar facilities. Be familiar with the department emergency plan criteria for transporting victims to more advanced medical care facilities.
- Know and follow procedures for protecting a potential crime scene.
- Know and follow department protocols for medical monitoring of response personnel involved or working with WMD and HAZMAT events.

HAZMAT

Address training requirements for HAZMAT team managers who will be part of the leadership and management of the emergency response team likely to respond to an event involving the terrorist/criminal use of WMD. These personnel will be involved in planning and managing the onsite scene involving the hot and warm zones. They will help set up the on-scene command post. HAZMAT team managers are expected to supervise staff who will attempt to mitigate and control the hazardous agents and materials. They are expected to use all available resources to bring the event to a successful conclusion. Generally, all of the actions to be taken by HAZMAT team managers should be conducted from within the warm zone or out to the cold zone. It is important that HAZMAT team managers are integrated into the overall command structure set up for management and supervision of resources and assets being deployed to control and conclude the WMD emergency event.

Planning and management level guidelines for HAZMAT team managers

- Have successfully completed training in awareness, performance, and management levels for events involving HAZMAT and WMD agents.
- Know and follow Incident Command System and Unified Command System procedures and requirements for implementing each system. Understand how the systems are implemented and integrated. Know what information the on-scene incident commander will need from the HAZMAT team manager. Be familiar with the full range of incident command functions and be able to fulfill any function pertaining to HAZMAT team operations.
- Know and follow protocols and procedures to secure, mitigate, and remove hazardous materials or potential WMD agents.
- Know and follow self-protection and protective measures for emergency responders to WMD events and hazardous materials events. Be aware of the special hazards to humans from WMD agents and hazardous materials.
- Know how to develop an incident action plan. Coordinate, with the on-scene incident commander, assets available for controlling WMD and hazardous materials events.
- Know and follow procedures for protecting a potential crime scene. Understand the roles and jurisdiction of federal agencies in a WMD event.
- Know and follow department protocols for medical monitoring of response personnel involved with or working onsite at WMD and HAZMAT events, including response team members involved with or working within the hot and warm control zones or personnel involved in onsite decontamination.

Emergency management

Address training requirements for emergency management personnel who will be part of the leadership and management team expected to respond to an event involving the terrorist criminal use of WMD. These emergency management directors/coordinators/team managers will be involved in preparing plans for mobilizing and coordinating the resources and assets needed for managing emergency operations and for providing onsite technical assistance when needed. These personnel will assist in planning implementation of the incident command structure, staffing of the satellite or headquarters emergency operations center, and establishing the command post or mobile command unit at the scene. These personnel are expected to work in a coordinated manner with the on-scene incident commander or unified command team to manage the emergency management resources required for bringing the event to a successful conclusion. Generally, many of the activities conducted by the emergency management team will be away from the immediate emergency incident scene, but require staying in communication with those at the scene. All of the actions to be taken by these emergency management team managers are expected to be conducted from within the cold zone.

It is expected that the emergency management team managers will be part of the overall response, but will not be part of the on-scene incident command structure, as called for in the emergency response preplan or the emergency operations plan document. However, the emergency manager is responsible for developing, testing, exercising, and revising the preplan or emergency operations plan established for coordinating management and supervision of the resources and assets that will be needed to control and successfully conclude the overall WMD emergency event.

Planning and management level guidelines for emergency management agency's emergency managers

- Have successfully completed appropriate and qualified training at the awareness and management levels for events involving HAZMAT and for WMD.
- Know and follow Incident Command System and Unified Command System procedures and requirements for implementing each system. Understand how the systems are implemented and integrated. Recognize when it is appropriate for the Unified Command System to evolve from the Incident Command System. Know what information the on-scene incident commander will need from the emergency management agency emergency operations center. Be familiar with the full range of coordinating activities and duties of the emergency management agency and all incident command functions. Assist

those persons who will be fulfilling functions related to the emergency operations plan.
- Know how to develop an Incident Action Plan and identify assets available for controlling WMD and HAZMAT events. Coordinate these activities with the on-scene incident commander. Be familiar with steps to take to assist in planning operational goals and objectives that are to be followed onsite in cooperation with the on-scene incident commander.
- Know and follow self-protection and protective measures for the public and for emergency responders to WMD events and hazardous materials events.
- Know and follow procedures for protecting a potential crime scene.
- Know how to interface with and integrate requisite emergency support services and resources among the emergency operations center (EOC) management and the incident or unified command on-scene incident management team. Be familiar with the coordination functions and procedures that are to be conducted by and with the EOC in support of on-scene emergency response activities.

Public works

Address training requirements for all public works supervisors, planners, engineers, and superintendent/agency directors employed by public works facilities associated with a local jurisdiction involved in planning for the emergency response to a WMD incident, including one at a public works facility. Public works facilities, such as wastewater treatment or drinking water operations or a nuclear power plant, may represent WMD targets within the local jurisdiction. Properly trained public works managers will improve the overall effectiveness of emergency planning and preparedness for response to an incident within the local jurisdiction. Typically, it is assumed that non-public works personnel will constitute the emergency response organizations or resources.

If, however, an incident were to occur at a power plant or other public works plant, the employees of the particular plant likely would be the first responders in protecting their own personnel.

Planning and management level guidelines for public works managers and supervisors

- Have successfully completed appropriate and qualified training at the awareness and management levels for events involving HAZMAT and WMD agents.
- Know and follow Incident Command System and Unified Command System procedures and requirements for implementing each system.

Understand how the systems are implemented and integrated. Know what information the on-scene incident commander will need from the public works supervisor or manager. Be familiar with the full range of coordinating activities and duties of the public works agencies. Understand the Incident Command System and the Unified Command System.
- Know how to develop appropriate plans for actions to be taken by the public works agency when a WMD and hazardous materials event occurs. Know how to coordinate plans with the on-scene incident commander. Know what steps to take to assist in planning operational goals and objectives that are to be followed onsite in cooperation with the on-scene incident commander in bringing the event to a successful conclusion.
- Know and follow self-protection and protective measures for the public and for public works emergency responders in WMD events and HAZMAT events.
- Know and follow procedures for protecting a potential crime scene.
- Know how to interface and integrate emergency support services and resources that will be needed (or are needed) among the EOC, the on-scene incident management team, and public works facilities and agencies. Be familiar with the coordination functions and procedures that are to be conducted by public works with the EOC to support on-scene emergency response activities.

First Responders' Resources

WMD civil support team (WMD-CST)

The WMD-CST was established to deploy rapidly to assist a local incident commander in determining the nature and extent of an attack or incident; provide expert technical advice on WMD response operations; and help identify and support the arrival of follow-on state and federal military response assets. They are joint units and, as such, can consist of both Army National Guard and Air National Guard personnel, with some of these units commanded by Air National Guard lieutenant colonels.

The mission of WMD-CST is to support local and state authorities at domestic WMD/nuclear, biological, and chemical (NBC) incident sites by identifying agents and substances, assessing current and projected consequences, advising on response measures, and assisting with requests for additional military support.

The WMD-CMT is able to deploy rapidly, assist local first responders in determining the nature of an attack, provide medical and technical advice, and pave the way for the identification and arrival of follow-on state and federal military response assets. They provide initial advice on what the agent may be and assist first responders in the detection assessment process. They are the first military responders on the ground, so that if additional federal resources are called into the situation, they can serve as an advance party that can liaise with the joint task force civil support.

The units provide critical protection to the force, from the pre-deployment phase of an operation at home station through redeployment. They ensure that strategic national interests are protected against any enemy, foreign or domestic, attempting to employ chemical, biological, or radiological weapons, regardless the level of WMD/NBC threat. They are a key element of the DoD's overall program to provide support to civil authorities in the event of an incident involving WMD in the United States.

They maintain the capability to mitigate the consequences of any WMD/NBC event, whether natural or man-made. They are experts in WMD effects and NBC defense operations.

These National Guard teams provide DoD's unique expertise and capabilities to assist state governors in preparing for and responding to chemical, biological, radiological, or nuclear (CBRN) incidents as part of a state's emergency response structure. Each team consists of 22 highly skilled, full-time National Guard members who are federally resourced, trained, and exercised, and employs federally approved CBRN response doctrine.

The WMD-CST is not designed to replace the first responder. The team integrates into the Incident Command System (ICS) in support of the local incident commander, providing a crucial capability between the initial local response and that of follow-on federal and state assets. Municipal fire, HAZMAT, police, and EMS agencies have a proven capability to deal with most emergencies. Larger incidents use mutual aid plans and the ICS to cope with the emergency. However, a WMD attack would present unique obstacles, such as identification of a military agent or spread of contamination, that could quickly overwhelm existing local and state resources. The WMD-CST provides rapid detection and analysis of chemical, biological, and radiological hazards agents at a WMD incident scene. The team is trained for CBRNE response and can provide advice on event mitigation, medical treatment, follow-on resources, and other response concerns to the incident commander.

WMD-CST key characteristics

- Must be certified by Secretary of Defense
- Unique to the National Guard
- Main role is support to governor and incident commander
- Operates only within U.S. territory
- Manned by Title 32 full-time (AGR) Army and Air National Guard personnel
- Interoperable with civil responders

CBRNE enhanced response force package (CERFP)

Another asset that the National Guard can offer in the near future is the CERFP. The NG CERFP consists of a core of full-time personnel augmented by traditional National Guard citizen-soldiers and airmen. The NG CERFP concept combines existing Army and Air Force National Guard medical, engineer, and security forces to leverage current force structure into a capabilities-based force packages with some adjustments in organization and equipment.

The CERFP can perform mass medical decontamination, technical casualty search and extraction, and emergency medical treatment in hostile WMD operating environments.

Functioning as a robust follow-on team capability to the WMD-CST, the NGCERFP is composed of five cells: command-and-control, security, medical, extraction and decontamination, and medical services. Much larger than the WMD-CST, the CERFP can operate for much longer durations. A CERFP typically consists of an enhanced division medical company with a decontamination and treatment capability and an enhanced engineer company. Currently, the teams are being formed and not all teams are functional.

Conclusion

"Weapon of mass destruction" is not a new term, and it is one used not only by arms-control specialists, but throughout history and the international diplomacy process. As one would expect of a term used in international agreements, it has an accepted meaning for use in disarmament negotiations and in defining treaty obligations accepted by the United States. Hence, any definition of the term WMD used as a matter of policy by the U.S. government should be consistent, effectively meaning nuclear, biological, and chemical, or chemical, biological, radiological, or nuclear.

chapter six

Understanding terror, terrorism, and their roots

Terrorism is as old as history itself. There have been acts of terrorism perpetrated against individuals, groups, and states in one form or another since individuals figured out how to use objects around them as weapons, and preventing the acquisition of a chemical, biological, or nuclear weapon ranks high on the list of priorities of every federal, state, and local law enforcement agency.

Terrorist operations can have one or more of these six general objectives:

1. *Physical*: Destroy something. The 9/11 attack sought to destroy those symbols of Western economic, military, and political power represented by the World Trade Center, the Pentagon, and potentially the White House or the U.S. Capitol building.
2. *Political*: Undermine authority. The incidents of kidnapping and subsequent beheadings of hostages in Iraq represent an attempt to influence a government or organization to do what a terrorist organization wants.
3. *Psychological*: Scare somebody or build morale. Although this represents a secondary effect of all objectives, it can be the primary objective. It can be both positive and negative.
4. *Personal*: Eliminate somebody. The assassination of key government figures or prominent individuals is an example of this objective.
5. *Profit*: Hold someone hostage or steal something. Most notably in Central and South America, terrorist organizations have kidnapped businessmen and held them for ransom.
6. *Revenge*: Retaliate against an individual or group. This was another objective of the 9/11 attack—to get revenge for actions taken or not taken by others.

Terrorism is a global issue, and threatens all those concerned. Those concerned are everyone, everywhere, in every nation around the globe. No one is safe from the acts of a terrorist, and the terrorist will strike anywhere, at any time, to cause chaos and mass panic in a country and its people, and to disrupt all normal social functions.

International terrorism has been a global issue for decades, and now is a major threat to the United States and its allies. The essence of terrorism motivated by political idealism its unanticipated, premeditated, intimidation by force and total disregard for human life.

The response to acts of terrorism has primarily been a law enforcement issue, and since 9/11 it has become the responsibility of the entire United States, U.S. government, and its allies around the globe. State, federal, and military authorities have responded to combat this issue. It is now up to all of us working together to fight this enemy.

Acts of terrorism are not limited to the use of military explosives, chemicals, biological agents, and radiological or nuclear materials. The use of common hazardous substances, hazardous waste, and illegal disposal are all potential weapons for the motivated terrorist. Telling the story of terrorism is both complex and simple, and this is my step-by-step explanation and overview of what terrorism is about.

The word "terrorism" first became popular during the French Revolution, when the *régime de la terreur* was initially viewed as a positive political tool that used fear to remind citizens of the necessity of virtue.

The use of violence to "educate" people about ideological issues has continued, but it has taken on decidedly negative connotations—and has become predominantly, though not exclusively, a tactic deployed by those who do not have the powers of state at their disposal. The study of terrorism and political violence presents a clear synopsis of some of the major historical trends in international terrorism.

Making careful distinctions between the motivations that drive political (or ethno-nationalist) terrorism and religious terrorism, and why the rise of religious terrorism coupled with the increased availability of weapons of mass destruction, may foretell an era of even greater violence. In the past, the main goal of the terrorist was not to kill, but to attract media attention to his cause in the hope of initiating reform.

"For the religious terrorist," however, "violence is first and foremost a sacramental act or divine duty executed in direct response to some theological demand or imperative." Religious terrorists see themselves not as components of a system worth preserving, but as "outsiders," seeking fundamental changes in the existing order. The bombing of the World Trade Center, the Oklahoma City attacks, and the sarin nerve gas attacks in Tokyo demonstrate that fundamentalists of any religious denomination are capable of extreme acts of terrorism.

What is terrorism?

No official, global definition of terrorism has been agreed upon, and definitions tend to rely heavily on who is doing the defining and for what purpose. Some definitions focus on terrorist tactics to define the term, while others focus on the actor. Yet others look at the context and ask if it is military or not.

Terrorism is a system of terror, and it can be intimidation by government, as directed by and carried out by the party in power. It is "a policy intended to strike with terror against those whom it is adopted for, and the employment of methods of intimidation to include the act of being terrorized." A terrorist is "anyone who attempts to further his views by a system of coercive intimidation as it applies to members of one extreme revolutionary society" (*Oxford English Dictionary*).

This is only one definition of what terrorism is, and after reading *Inside Terrorism* by Bruce Hoffman, there are so many different descriptions of what terrorism is about. Over the last two centuries the term has changed and been applied differently, depending on whom is using the term.

So what is terrorism? Well, I give my view of what terrorism is, and continue from this perspective, building a step-by-step picture of what it is.

Terrorism is the use of force or violence against persons or property for the purpose of intimidation, coercion, or ransom. Terrorists often use violence and threats to create fear among the public, to try to convince people that their government is powerless to prevent acts of terrorism, and to get immediate publicity for their causes.

Acts of terrorism can range from threats to actual assassinations, kidnappings, airline hijackings, bomb scares, car bombs, building explosions, mailings of dangerous materials, agro terrorism, computer-based attacks, and the use of chemical, biological, and nuclear weapons—weapons of mass destruction (WMD) ("Terrorism", *Talking About Disaster: Guide for Standard Messages*, March 2007).

Why talk about terrorism?

In addition to the natural and technological hazards described elsewhere in this guide, people face threats of terrorism posed by extremist groups, individuals, and hostile governments. Terrorists can be domestic or foreign, and their threats to people, communities, and the nation range from isolated acts of terrorism to acts of war.

High-risk terrorism targets include military and civilian government facilities, international airports, large cities, and high-profile landmarks. Terrorists might also target large public gatherings, water and food supplies, utilities, and corporate centers. They are capable of spreading fear by sending explosives or chemical and biological agents through the mail.

The commonly accepted meaning of the word terrorism is any use of terror in the form of violence or threats meant to coerce an individual, group, or entity to act in a manner in which any person or group could not otherwise lawfully force them to act. The world community has struggled with creating a legal definition of terrorism that is globally accepted. In the United States, terrorism is broken down into two categories.

The first category is international terrorism, and the second is domestic terrorism. The main characteristics of acts or threats that constitute terrorism are the same as in the wider definition above, but additional details have been added. Both categories include violent acts or any actions that endanger human life or violate U.S. laws, both federal and state. The definition further clarifies that these acts appear to be intended to intimidate or coerce civilians—collectively—or to intimidate or coerce changes in governmental policy or government conduct through the use of mass destruction, assassination, or kidnapping.

The main difference between the two definitions of terrorism is the inclusion under international terrorism that clarifies criminal violations as those that would be construed as such if committed in U.S. federal or state jurisdictions. The other distinction describes where acts of terrorism occur. International terrorism occurs primarily outside the "territorial jurisdiction" of the United States, and domestic terrorism primarily occurs within the U.S. jurisdiction.

Despite legal definitions, terrorism means different things to different people. While the threat to the modern world appears to involve more Islamist extremists than any other type of terrorist, it should be clearly acknowledged that not all Muslims are terrorists and not all terrorists are Muslim. In fact, many Muslims are in as much danger under the radicalized practices of Islam as any other people, because they divert from the extremist interpretation of Islam.

Terrorism has been around for centuries, although it's been defined or described differently throughout the years. Today, it amounts to bullying, only on a grand scale, because these bullies have powerful weapons and fighters who embrace death as martyrdom. There is nothing more dangerous than a terrorist who believes that he has nothing to lose (from "What is terrorism?" by Sherry Holetzky).

Official U.S. government definition of terrorism

"[An] act of terrorism, means any activity that (A) involves a violent act or an act dangerous to human life that is a violation of the criminal laws of the United States or any State, or that would be a criminal violation if committed within the jurisdiction of the United States or of any State; and (B) appears to be intended (i) to intimidate or coerce a civilian population; (ii) to influence the policy of a government by intimidation or coercion; or (iii) to affect the conduct of a government by assassination or kidnapping."

(*United States Code Congressional and Administrative News, 98th Congress, Second Session*, 1984, Oct. 19, volume 2; par. 3077, 98 STAT. 2707 [West Publishing Co., 1984])

Looking at all of the terms that describe terrorism, the United States and other countries each have a slightly different twist on the term. Looking at the U.S. government and all of its agencies, they all have their own definition of what terrorism is, depending on what their response or responsibility is.

General history of terrorism

The history of terrorism is as old as humans' willingness to use violence to affect politics. The Sicarii were a first-century Jewish group who murdered enemies and collaborators in their campaign to oust their Roman rulers from Judea.

The Hashhashin, whose name gave us the English word "assassins," were a secretive Islamic sect active in Iran and Syria from the eleventh to the thirteenth century. Their dramatically executed assassinations of political figures terrified their contemporaries.

Zealots and assassins were not, however, really terrorists in the modern sense. Terrorism is best thought of as a modern phenomenon. Its characteristics flow from the international system of nation-states, and its success depends on the existence of a mass media to create an aura of terror among many people (http://terrorism.about.com/od/originshistory/).

The origins of modern terrorism

The word "terrorism" comes from the Reign of Terror, instigated by Maximilien Robespierre in 1793, following the French Revolution. Robespierre, one of twelve heads of the new state, had enemies of the revolution killed, and installed a dictatorship to stabilize the country. He justified his methods as necessary in the transformation of the monarchy to a liberal democracy: Subdue by terror the enemies of liberty, and you will be right, as founders of the Republic.

Robespierre's sentiment laid the foundations for modern terrorists, who believe violence will usher in a better system. But the characterization of terrorism as a state action faded, while the idea of terrorism as an attack against an existing political order became more prominent (http://terrorism.about.com/od/originshistory/).

The rise of non-state terrorism

The rise of guerrilla tactics by non-state actors in the last half of the twentieth century was due to several factors. These included the flowering of ethnic nationalism, anti-colonial sentiments in the vast British, French, and other empires, and new ideologies, such as communism.

Terrorist groups with a nationalist agenda have formed in every part of the world. They use suicide bombing and other lethal tactics to wage a battle for independence against the majority government.

Terrorism turns international

International terrorism became a prominent issue in the late 1960s, when hijacking became a favored tactic. The era also gave us our contemporary sense of terrorism as highly theatrical, symbolic acts of violence by organized groups with specific political grievances. They used spectacular tactics to bring international attention to their national cause.

The twenty-first century: Religious terrorism and beyond

Religiously motivated terrorism is considered the most alarming terrorist threat today. Groups that justify their violence on Islamic grounds—al-Qaeda, Hamas, Hezbollah—come to mind first. But Christianity, Judaism, Hinduism, and other religions have given rise to their own forms of militant extremism. They are not simply orthodox believers turned violent, but rather violent extremists who manipulate religious concepts for their own purposes.

No two terrorist groups are alike. Some groups operate worldwide, while others are regional. They fight for different reasons, with a variety of weapons and targets. As these terrorist organizations wage war and the Jihad continues the global war on terrorism, there will still be a concern about the threat and use of biological, chemical, nuclear, and radiological weapons. The main purpose of those terrorist groups who would launch such attacks is to strike fear and panic in millions.

Violence in the Koran and the Bible

What motivates terrorists to blow themselves up with explosives in public places in order to kill the largest possible number of innocent people? The impetus for such heinous crimes must be very powerful, and these senseless acts of terrorism have changed our lives forever.

In seeking to find a solution to the problem of Muslim terrorism, it is important to understand what motivates these people to engage in this frightening self-destruction. We are told by many moderate Muslims and political leaders that Islam is a religion of peace and it does not allow the killing of innocent people. The deplorable terroristic acts we have witnessed in recent times are supposed to be condemned by the teachings of Islam.

I do not want to judge Islam as a violent religion on the basis of some terroristic acts committed by those who claim to follow its teachings, and the same must be said of Christianity. We cannot conclude that Christianity teaches violence because of the violent crusades some Christians waged in the past against Muslims, Jews, and so-called "heretics." Such an interpretation is wrong, because not all who claim to act in the name of their religion are necessarily following its teachings.

The correct approach is to go back to the sources of Islam and Christianity and see what they have to say about violence and peaceful coexistence with people practicing other religions. This is the procedure we shall follow by examining, first, what Islam has to say about warfare, and then, by comparing its teachings with those of the Bible.

The teaching of Islam about Jihad or fighting for the cause of Allah

"Some scholars view as futile the attempt to define the teaching of the Koran and the Hadith (collected teachings of Muhammad) regarding the use of warfare to advance the cause of Allah. The Koran is a vast, vague book, filled with poetry and contradictions (much like the Bible). You can find in it condemnations of war and incitements to struggle, beautiful expressions of tolerance and stern pictures against unbelievers."

Is it true that the Koran is a contradictory book that condemns war on the one hand and commands warfare on the other hand? The answer is "No!" The contradictions in the Koran are resolved by recognizing Muhammad's progressive teachings from peace to war during the course of his life and experiences.

Muhammad ordered the waging of war against those who attacked him, which happened later in his life as he was gathering followers of Islam. I looked up and read many chapters of the Koran and came to the conclusion that attempting to understand it was bewildering. Like the Bible that I read and follow, anyone can take its teaching and turn it into whatever they want to twist out of it. So I use what I want to get what I want out of it to justify my actions. This has happened over the centuries and continues to the present day. *Violence in the Koran and the Bible, Dr. Samuele Bacchiocchi, June 07, 2002*

What is Islam?

Before continuing, I had to look up what Islam was and learn some of its teachings. To my surprise, it looks really similar to what I learned as a child about the teachings in the Holy Bible.

The literal meaning of Islam is peace; surrender of one's will, losing oneself for the sake of God, and surrendering one's own pleasure for the pleasure of God. The message of Islam was revealed to the Holy Prophet Muhammad (peace and blessings on him) 1400 years ago. It was revealed through the Angel Gabriel (on whom be peace) and was thus preserved in the Holy Quran. The Holy Quran carries a Divine guarantee of safeguard from interpolation and it claims that it combines the best features of the earlier scriptures.

The prime message of Islam is the Unity of God; that the Creator of the world is One, and He alone is worthy of worship, and that Muhammad (peace and blessings on him) is His Messenger and Servant. The follower of this belief is thus a Muslim. A Muslim's other beliefs are God's angels; previously revealed Books of God; all the prophets, from Adam to Jesus (peace is on them both); the Day of Judgment; and indeed the Decree of God. A Muslim has five main duties to perform, namely; bearing witness to the Unity of God and Muhammad (peace and blessings on him) as His Messenger, observing the prescribed prayer, payment of Zakat, keeping the fasts of Ramadhan, and performing the pilgrimage to Mecca.

Islam believes that each person is born pure. The Holy Quran tells us that God has given human beings a choice between good and evil, and to seek God's pleasure through faith, prayer and charity. Islam believes that God created mankind in His image, and by imbuing the attributes of God on a human level, mankind can attain His nearness. Islam's main message is to worship God and to treat all God's creation with kindness and compassion. Rights of parents in old age, orphans, and the needy are clearly stated. Women's rights were safeguarded 1400 years ago, when the rest of the world was in total darkness about emancipation. Islamic teachings encompass every imaginable situation, and its rules and principles are truly universal and have stood the test of time.

In Islam, virtue does not connote forsaking the bounties of nature that are lawful. On the contrary, one is encouraged to lead a healthy, active life with the qualities of kindness, chastity, honesty, mercy, courage, patience, and politeness. In short, Islam has a perfect and complete code for the guidance of individuals and communities alike. As the entire message of Islam is derived from the Holy Quran, and indeed the Sunnah and Hadith (the traditions and practices of the Holy Prophet, peace and blessings on him), it is immutable in the face of changes in time and place. It may appear rigid to the casual eye; in actual fact, it is most certainly an adaptable way of life, regardless of human changes.

Islamic radicals

Radical Islam is a militant, politically activist ideology whose ultimate goal is to create a worldwide community, or caliphate, of Muslim believers.

Determined to achieve this new world order by any means necessary, including violence and mass murder, radical Islam is characterized by its contempt for the beliefs, practices, and symbols of other religious traditions. This intolerant creed is cited by Islamists as the philosophical justification for their terrorism.

Radical Islam's kinship with terrorism, and its willingness to use violence as a means to its ultimate ends, is clearly spelled out in a training manual produced by the radical Islamist terror group al-Qaeda, whose operatives inflict destructive terror upon whatever target they deem a threat to their ideologies.

The issue with using verses from the Koran and the Bible is that anyone can misinterpret statements in these texts as to what one must do to follow the faith. Violence is the only way, and anyone not understanding this view is an infidel and must be destroyed. Those that are not followers are non-believers of the faith.

Although many Muslims have strong feelings about their heritage, many have mixed feelings about their current political status. Nearly all have a strong sense for their religion and its contribution to civilization.

With this said, terrorist organizations can take advantage of those seeking someone to follow because of social need, personal conflict with themselves, or outside influences directly impacting their immediate social circle.

The inner circle of the Muslim world is unique in how it is set up and the loyalties that are involved with this social structure. There is strict, hierarchical order in a Muslim's world.

The order of allegiance is extremely important: listed below is the order of allegiance.

- Family (kham)
- House (beit)
- Clan (fukhdh)
- Tribe (qabila)
- Region (e.g. Kurdistan)
- Nation (e.g. Iraq, Afghanistan)
- Ethnicity (e.g. Arab, Kurd)
- Sect (e.g. Sunni, Shia)
- Religion (e.g. Islam)
- People of the book (e.g. Muslim, Jewish, Christian)
- Infidels

This hierarchy of allegiance is obligatory, and it is followed for the following reasons. The tribal affiliation is necessary for survival, and the large extended family has the clans and households within its structure.

Within these, loyalties are always to the immediate family first. Decisions are based on the family's position in society, and the tribes make and enforce their own laws.

Importantly, tribal bonds diminish in the cities, so the loyalties are more to oneself or local surrounding friends or groups. This is where it becomes much easier to recruit for members of these terrorist organizations. Last on the list are the infidels, which can be anyone, including Muslims who are considered weak or no-faith non-believers, or who do not believe in following the current leader or their order of things.

Depending on where you sit in the hierarchy, you may have no power to change anything around you to improve your social status within these groups. Many times, it's "do as you're told"—without any input, and the social order remains (Figure 6.1).

This chart shows the hierarchy of allegiance. Information for this structure comes from ROC-IED ABCA U.S. Army training course 2011

The chart shows how allegiance is used and followed in Muslim society. It is strictly followed, and the rules are enforced by the family and the elder male in the family. This chart which shows the hierarchy of importance. The further away from the family, the less value you have to the family structure hierarchy. The trust and confidence factor score reflects the level of control and importance one has in these

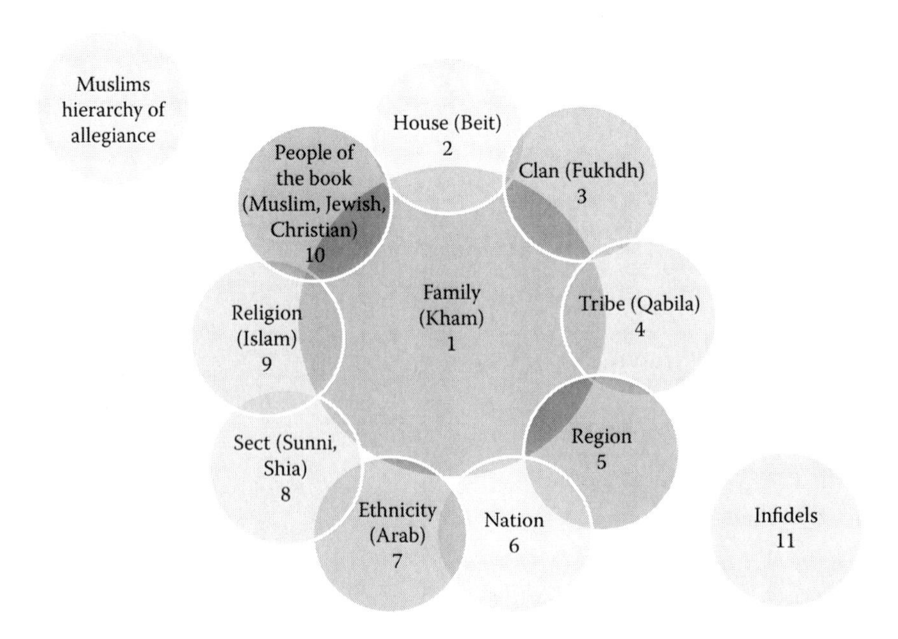

Figure 6.1 Hierarchy of allegiance.

operative groups, with values less than one reflecting lower levels of influence.

Terrorist tactics

The attacks against governments, military, and commercial targets typically take the following forms:

- Attacks on transportation systems, buildings, and other soft targets using improvised explosive devices.
- Ambushes on police, security forces, and the general population are also common.
- Sabotage of oil pipelines and other infrastructure is another tactic often used.
- Assassination of political figures cooperating with outside governments, such the United States or coalition forces in the Middle East
- Suicide bombings targeting international organizations, police, hotels, and so on.
- Kidnapping and murder of private contractors working in the Middle East for commercial entities.
- Kidnapping private citizens as a fundraising tactic.

Improvised explosive devices

The majority of terrorist attacks come in the form of IEDs targeting soft targets. Most IEDs are made from leftover munitions and foreign explosive materials which are often hastily put together. Vehicle-borne IEDs are devices that use a vehicle as the package or container for the device. These IEDs come in all shapes and sizes, from small sedans to large cargo trucks.

There have even been instances of what appeared to be generators, donkey-drawn carts, and ambulances used to attempt attacks on people in public gathering places and government entities.

The internet and psychological warfare

Terrorism has often been conceptualized as a form of psychological warfare, and terrorists have certainly sought to wage such a campaign through the internet. There are several ways for terrorists to do so. They can use the internet to spread disinformation, to deliver threats intended to instill fear and helplessness, and to disseminate horrific images of recent actions, such as beheadings of foreign hostages in Iraq. Terrorists wage battles using traditional guerrilla tactics, as well as employing psychological warfare on the internet. Many terrorist groups use message boards, online chat, and religious justifications for their activities. Sites also often provide histories of their host organizations and activities.

The goal of these terrorism tactics is to change government policies or the structure of a government: terrorists seek to embarrass and undermine the leadership with their activities. Inflicting random attacks like shootings, bombings or kidnappings, the terrorists seek to erode the government's credibility and influence.

They want to show citizens that their government is incapable of maintaining public safety and order, leading to public outcry for government officials to cede to terrorist demands in order to stop the violence and restore stability.

Terrorist training camps

There are, and have been, training camps for terrorists. The range of training depends greatly on the level of support the terrorist organization receives from various organizations and states. In nearly every case, the training incorporates the philosophy and agenda of the group's leadership as justification for the training, as well for potential acts of terrorism which may be committed. State-sanctioned training is by far the most extensive and thorough, often employing professional soldiers and covert operatives of the supporting state.

Terrorist incident activities

The following chart shows terrorist attacks from 1988 to 2010, including the types of weapons used during those attacks (Figure 6.2).

This data comes from a variety of sources, including the Global Terrorist Database, CENTCOM, U.S. Army/terrorism/Intelligent.

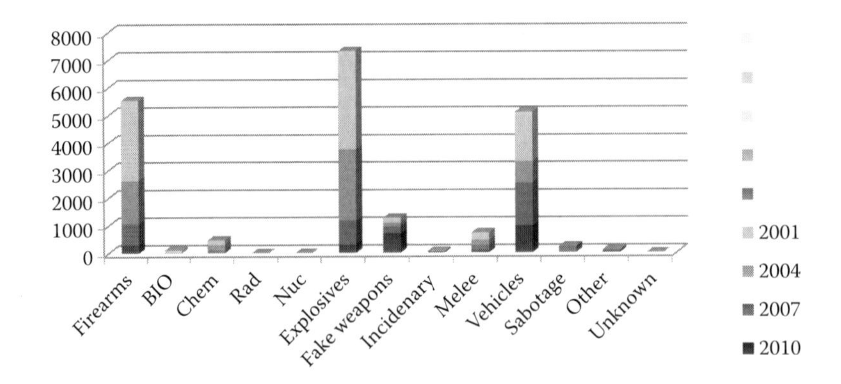

Figure 6.2 Terrorist incident activities.

This data only shows official attacks and the types of weapons used during those attacks. As one can see, there are a vast variety of weapons used, and many are easily available. Some are more difficult to get, in that many of the attacks involve the use of explosives and small-arm weapons. This information comes from variety of sources, and is intended only to show how frequently these attacks occur and the weapons used during these attacks.

Whether these attacks are individual or group attacks, small or large, they all cause mass chaos and financial cost to the country and place fear in the people.

The cost of terrorism

The United States alone now spends about US$500 billion annually—20% of the U.S. federal budget—on departments directly engaged in combating or preventing terrorism, most notably Defense and Homeland Security. The defense budget increased by one-third, or over $100 billion, from 2001 to 2003, in response to the heightened sense of the threat of terrorism.

- The increased risk and prevalence of global terrorism looms as a major threat to all countries. Terrorist acts have already imposed significantly increased costs on all economies.
- The immediate costs of terrorist acts, including loss of life, destruction of property, and depression of short-term economic activity, are compounded by the costs associated with the continuing threat of terrorism.
- Terrorism unchecked creates uncertainty, reduces confidence, and increases risk perceptions and risk to economic growth. Terrorist acts can severely disrupt international trade, tourism, and the continuing threat of terrorism imposes an even greater rise in these costs and reduces revenue, thus having a global effect.
- Given their greater reliance on trade and capital inflows, developing APEC economies may incur higher costs, relative to GDP, from unchecked terrorism.

Effective action to combat terrorism will generate significant benefits for the global economy, preventing losses from reduced trade flows and investment, which undermine economic growth. The failure to counter terrorism will produce costs for all economies and populations.

During the past 35 years, there have been nearly 20,000 terrorist incidents (Figure 6.3).

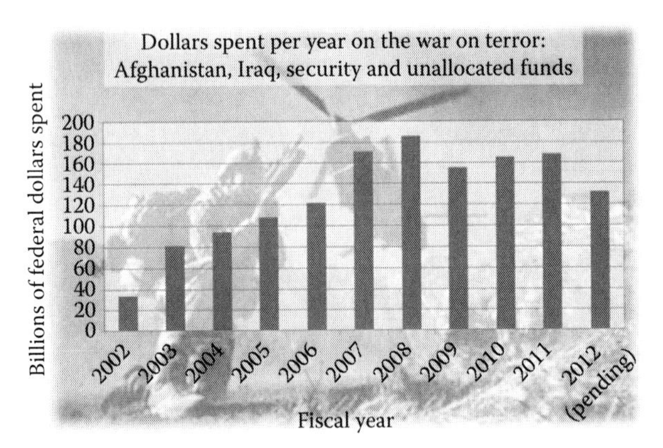

U.S. congressional research service, March 29, 2011 report United States alone

Figure 6.3 Defining terrorism.

The projected cost for the war on terrorism is estimated around 6.5 trillion dollars. Half of this cost belongs to the United States, and the other half will come from the rest of those countries combating terrorism. A staggering cost will arise, and for this the reason the global community needs to keep up the war on terrorists, taking away their ability to wage this Jihad on the global community (Figure 6.4).

Bath, J., Economic Impacts of Global Terrorism: From Munich to Bali October 2006;

Reyko Huang, Terrorism Project, Dec 20, 2001t

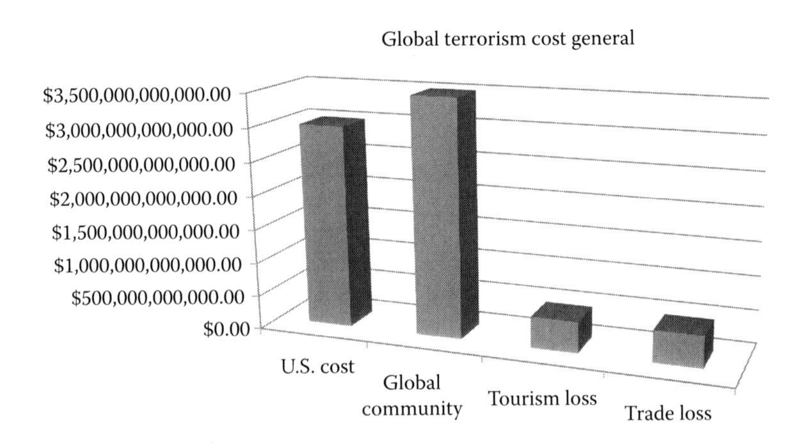

Figure 6.4 Global terrorism cost.

This data above shows the financial impact the war on terrorism has had on the international community. These costs are projected to rise in the future.

The overall cost for waging the war on terrorism is staggering; the cost figures are only estimated, and could rise in the future. These numbers illustrate the huge amount of money that is being poured into this war. There are many other costs associated with fighting and conducting this war.

Conclusion

Terrorism is a complex problem with many diverse causes. Consequently, no single effective method to counter it exists. To combat terrorism, one must first understand the underlying motivations for each particular group's actions. A strategy then needs to be developed based on those findings. Regardless, it is difficult to fight terrorism without endangering civil liberties, as seen in Northern Ireland. Many innocent people get caught in the cross-fire. Ending terrorist threats requires imaginative and fluid thinking, whether to attack the roots of terrorism or neutralize a particular group.

According to terrorism expert Paul Wilkinson, the general principles which have the best track record in reducing terrorism include

- No surrender to the terrorists, and an absolute determination to defeat terrorism within the framework of the rule of law and the democratic process.
- No deals and no concessions, even in the face of the most severe intimidation and blackmail.
- An intensified effort to bring terrorists to justice by prosecution and conviction before courts of law.
- Tough measures to penalize the state sponsors who give terrorist movements safe haven, explosives, cash, and moral and diplomatic support.
- A determination never to allow terrorist intimidation to block or derail international diplomatic efforts to resolve major political conflicts in strife-torn regions, such as the Middle East. In many such areas terrorism has become a major threat to peace and stability, and its suppression therefore is in the common interests of international society.

chapter seven

CBRNE and weapons of mass destruction

What are weapons of mass destruction? They are chemical, biological, radiological, and nuclear (CBRN) weapons that many countries have and would use as a means of defense or terrorism. These weapons have the capacity to inflict death and destruction on such a massive scale and so indiscriminately that their very presence in the hands of a terrorist, criminal, or hostile power can be considered a grievous threat. Modern weapons of mass destruction are nuclear, biological, or chemical weapons—frequently referred to collectively as NBC weapons.

There are rogue countries that would use these weapons as a means to an end, and these rogue states also fund, train, supply, and equip terrorist groups around the globe to back their extremist agenda, especially against countries in the West; the United States, and Western Europe.

Chemical weapons (CW)

Chemical weapons (CW) are defined as weapons using the toxic properties of chemical substances rather than their explosive properties to produce physical or physiological effects on an enemy.

Attributes of CW:

- CW are relatively inexpensive to produce
- CW can affect opposing forces without damaging infrastructure
- CW can be psychologically devastating
- Blister agents create casualties requiring attention and inhibiting force efficiency
- Defensive measures can be taken to negate the effect of CW
- Donning of protective gear reduces combat efficiency of troops
- Key to employment is dissemination and dispersion of agents
- CW are highly susceptible to environmental effects (temperature, winds)
- Offensive use of CW complicates command-and-control and logistics problems

Chemical weapons include such agents as mustard, tabun, sarin (GB), and nerve gas (G- and V-agents). VX gas is a V-type nerve gas. In the 1960s, considerable work went into developing CS gas as a quick-acting, non-lethal tool for controlling riots and civil disorder. Later, a class of non-lethal, mental incapacitants causing mental disorientation were developed, most notably BZ agents.

Cyanides (sodium, potassium, or hydrogen) are better described as poisons than CW agents. They must be disseminated in high concentrations, or in enclosed and unventilated areas to be effective.

Relatively large quantities of CW are needed to achieve mass casualties. According to a pentagon estimate, 22 lb of sarin would be needed to kill 50 people in an outdoor environment. It would take about 220 lb to kill 500, and 2,200 lb to produce 10,000 casualties. Naturally, less would be needed in an enclosed space. Although this level of casualties is higher than that from conventional explosives, the problems of collecting and disseminating sufficient quantities are formidable. The efficiency of disseminating the agent is critical to the effectiveness of CW. The methodology depends on whether the agent is designed as an aerosol, liquid, or solid and how it is to be absorbed by the target. Weaponized CW agents are often disseminated by explosives, such as artillery shells.

While tactical chemical weapons can be used in a battlefield environment, they are inappropriate and problematic for use by terrorists as weapons of mass destruction. To qualify as WMDs, weapons must be able to produce mass casualties.

This would require large quantities of chemicals, combined with an effective dispersion system. It is simply impractical for terrorists to secretly obtain, transport, and disperse the large quantities of chemicals required to attain a significant result.

Chemical weapons have a relatively small area of influence and quickly disperse into the air or settle to the ground. When combined with explosives to increase dispersion, the active chemicals are often destroyed or degraded by the explosive blast. Terrorist groups have extensive experience with conventional explosives and gain little advantage from chemical weaponry.

Many countries have the ability to produce, or have produced chemical weapons, and terrorist groups could acquire or produce such materials. In Japan, the Aum Shinrikyo cult used sarin gas in an attack in a Tokyo subway. Although this incident is often cited to alert people to the threat of chemical weapons, only 11 people died from the attack, which was carried out in an enclosed area.

The terrorist threat from chemical weapons is overstated. The sheer volume of chemicals required to have a significant impact is a major barrier to the use of CW. It would be difficult to acquire, produce, or transport chemical agents in sufficient quantities without attracting attention.

Equally challenging is managing dispersal effectively and to do so without impacting on those releasing the chemical agents.

Many common toxic industrial chemicals could be used as weapons. What they lack in toxicity is made up by the large quantities commonly available and accessible. Chlorine and phosgene gases are industrial chemicals that are regularly transported in bulk road and rail shipments. Terrorist/criminal acts could easily target commercial containers and rupture them to release the gases. The effects of chlorine and phosgene are similar to those of mustard agent. Chlorine and other chemical spills from trucks and railcars are not uncommon; terrorists would simply need to select targets and timing to maximize the effects on the public. Trucks and railcars are notoriously vulnerable targets to which little attention has been directed.

Security measures must be improved to protect trucks and railcars passing through, or parked in, population centers. Similar precautions are needed where ships are off-loading bulk chemicals.

In December 1984, an explosion at the Union Carbide pesticide factory in Bhopal, India, released methyl isocyanate (MIC), a highly toxic gas. MIC, hydrogen cyanide, and at least 65 other gases spread across the city in a cloud, killing over 5000 people within 3 days. At least 20,000 people have died as a result of exposure to the gases. This incident demonstrates the lethal effect of toxic gases, and the large quantities needed to produce a disastrous impact. It should also serve as a reminder of the need to improve security at chemical plants where terrorists could seek to replicate the Bhopal incident.

Another point of vulnerability exists in building heating, ventilating, and air conditioning (HVAC) systems. Building HVAC typically recirculates air from the occupied space to the air-handling units and back again. The systems add a small percentage of outdoor, make-up air to replace that lost through exfiltration. It is possible to deliver a chemical or biological agent at the point where outdoor air is drawn into the building system, and this can be done without entering the building. The air handlers will disseminate the toxic agent throughout the building. Air intakes are often difficult to access, but those that are not must be adequately protected.

Chemical or biological agents could also be introduced through return air inlets in virtually any building where terrorists had access, including many apartments, commercial offices, or public facilities (sports stadiums, theaters, etc.) As tenants, terrorists would have time and access to build up sufficient inventories of CW and BW agents to maximize the effect. Building owners and operators must be alerted to these possibilities and encouraged to implement measures to deny access to air intake points. *High value target buildings, with highly vulnerable HVAC systems, must receive special attention.*

Biological weapons (BW)

A biological weapon disperses organisms, or micro-organisms, to produce disease in humans, plants, and animals. The mortality rates vary among the various diseases. The most dangerous are those that are communicable and can be passed from one infected victim to others, but not all are communicable. Anthrax and ricin are nor communicable; smallpox is.

While chemical weapons have been used for decades, biological weapons are a relatively new development because the production and weaponization of biological agents is far more difficult and sophisticated, and has been made possible by technological advances during the late twentieth century.

Given the complexities in producing, managing, and disseminating biological agents effectively to maximize fatalities, BW is probably the least feasible approach for large-scale terrorist attacks. On a smaller scale, BW becomes a more attractive option, but still entails considerable technical prowess that is not required for conventional explosives.

The anthrax scares following the terrorist attacks of 9/11 cast attention on the emerging threat of biological warfare. The threats most commonly mentioned are from anthrax, botulism, ricin and smallpox.

Anthrax—Bacillus anthracis, the bacterium that causes anthrax, is capable of causing mass casualties. Symptoms usually appear within 1–6 days after exposure and include fever, malaise, fatigue, and shortness of breath. The disease is usually fatal unless antibiotic treatment is started within hours of inhaling anthrax spores; however, it is not contagious.

Few people are vaccinated against anthrax. Anthrax can be disseminated in an aerosol or used to contaminate food and water; when ingested, mortality rates are highest. Anthrax can be contracted by skin contact, in which instance it is rarely fatal.

Botulinum toxin—occurs naturally in the soil. Crude but viable methods to produce small quantities of this lethal toxin have been found in terrorist training manuals. Symptoms usually occur 24–36 h after exposure, but onset of illness may take several days if the toxin is present in low doses. Botulinum toxin would be effective in small-scale poisonings or aerosol attacks in enclosed spaces, such as movie theaters. The toxin molecule is likely too large to penetrate healthy skin.

Ricin—a plant toxin that is 30 times more potent than the nerve agent VX by weight and is readily obtainable by extraction from common castor beans. There is no treatment for ricin poisoning after it has entered the bloodstream. Victims start to show symptoms within hours to days after exposure, depending on the dosage and route of administration. Terrorists have looked at delivering ricin in foods and as a contact poison, although it is not known to penetrate healthy skin. Ricin remains stable in unheated foods and has few indicators because it does not have a strong taste or color.

Smallpox—has been eradicated as a result of a successful worldwide vaccination program. Only two stockpiles of the variola virus are known to exist; in Russia and the United States. Generally, direct and fairly prolonged face-to-face contact is required to spread smallpox from one person to another. Smallpox also can be spread through direct contact with infected bodily fluids or contaminated objects such as bedding or clothing.

Rarely, smallpox has been spread by virus carried in the air in enclosed settings such as buildings, buses, and trains. Humans are the only natural hosts of variola. Smallpox is not known to be transmitted by insects or animals. Smallpox is small enough to be inhaled, so it could be spread in an aerosol. The virus is very stable, which means it is not easy to destroy, and it retains its potency for days outside a human host.

Since there is no cure for smallpox, the only way to deal with the disease is by vaccination. Since the variola virus is essentially extinct, it would be extremely difficult to produce, and because of its lethality, stockpiles are unusually secure. Terrorists would face enormous technical challenges to produce weaponized smallpox virus and would be unable to control the direction an outbreak would take, meaning it could be turned against their own people.

Pathogen feed stocks

There are four sources of pathogen feed stocks: (1) natural sources; (2) culture collections; (3) research laboratories and public health facilities; and (4) state sponsors of BW programs.

Developing biological agents from scratch (natural sources) is a significant challenge, requiring sophistication, financing, and considerable luck to produce quantities needed for mass destruction. Experts suggest that producing small quantities of some biologicals is reasonably feasible, even by unsophisticated methods. However, fears of large-scale attacks appear to be an over-reaction.

Since 9/11, intelligence and police forces have become far more aware and vigilant to signs of attempts to produce WMDs of any sort, including identification of purchases of related materials and the security of existing stockpiles.

The United States and Germany have passed laws regulating the transfer of pathogens from laboratories and commercial firms, but many countries have not.

Encouraging nations to adopt laws and international regulations on the transfer of biologicals would make development and proliferation of biologicals significantly more difficult at little cost.

BW delivery systems

Most biologicals are not communicable, which means the agent must be delivered to each target individual. Biological agents are susceptible to adverse effects from improper storage, temperatures, humidity or exposure to sunlight, oxygen, or other materials.

Most have limited shelf life and lose their potency over time. Given the lethal effects of BW agents, they require extremely careful handling.

BW agents begin in a liquid form, which presents technical challenges to effective dissemination methods. The active agents must be also being reduced to a specific size in the range 1–5 microns, requiring elaborate and difficult procedures. Converting the materials to dry form offers a number of advantages, but even greater technical prowess and equipment.

The processes are complex and fraught with potential for failure. These are not the kind of tactics favored by terrorist, who prefer simple plans, cleverly and flawlessly executed. One does not simply hijack a crop duster and pour BW agents into the tank as a substitute for pesticides.

Notwithstanding the heinous attacks of 9/11, even the worst terrorists must consider the consequences of their acts. While any attack is designed to hurt and weaken the enemy, the primary goal is to expand the base of support for the movement. No nation has ever used biological warfare, and being the first to re-introduce an extinct disease to mankind is of dubious propaganda value.

The fact is that a BW attack is unnecessary—WMDs are in the news daily, and Americans are already reacting with frenzied spending to upgrade homeland security against every conceivable form of attack.

Radiological weapons (RW)

Radiological dispersal devices (RDDs), also called "dirty bombs," combine a conventional explosive with some form of radioactive material. Such an improvised device does not produce a chain reaction or nuclear detonation; it merely uses the explosive to spread radioactive material across a localized area. Ideally, for the terrorist, the winds will help spread the radioactivity to a larger area.

When press reports mention a "backpack bomb," they're referring to a small, improvised RDD that includes a small explosive charge and radiological material in a shielded tin can called a "pig." It's improbable that such a device could produce a nuclear detonation, as some press accounts have suggested. A backpack bomb would probably not even qualify as a weapon of mass destruction.

The CIA reports that "A variety of radioactive materials is commonly available and could be used in an RDD, including Cesium-137,

Strontium-90, and Cobalt-60. Hospitals, universities, factories, construction companies, and laboratories are possible sources for these radioactive materials."

The Department of Homeland Security needs to require increased security measures at all sites and facilities that store and utilize radiological materials, while the EPA should enact new regulations addressing such materials.

An RDD detonated by terrorists would cause casualties in the immediate area as well as health, environmental, and economic effects. Its greatest impact would undoubtedly be the psychological effect—the terror. Depending on the amount of radioactive material released, the size of the area affected, and land levels of contamination, such an incident could force closure of the affected area for a long period to conduct expensive cleanup and remediation work.

Despite the expert analysis and predictions regarding potential terrorist targets, none have mentioned the most vulnerable, highest-risk radiological target, likely to attract the attention of terrorists—nuclear waste shipments.

The U.S. Department of Energy ships tons of transuranic radioactive waste every week from national laboratories and other nuclear sites across the country to the Waste Isolation Pilot Plant (WIPP) in Carlsbad, New Mexico. The "hot" materials are stored in heavy-duty casks, designed to withstand truck crashes, but not explosives. The entire transportation security system was designed with little consideration of potential terrorist attacks. The flat-bed trailers with two or three huge casks are easily identifiable.

The truck fleet follows the same routes along major interstate highways, with dozens of trucks and tons of transuranic wastes on the road at any one time. The published routes funnel the trucks into several choke points in several major metropolitan areas, namely Denver, Dallas, and Albuquerque. A coordinated terrorist attack, timed to coincide with optimum wind conditions, could easily use an improvised explosive device (IED), like those employed in Iraq, or a car or truck bomb to rupture the casks and disseminate radioactivity across several large cities.

Al-Qaeda has already demonstrated a proclivity for nearly simultaneous attacks against remote targets on 9/11 and in the bombings in Kenya and Tanzania. This is precisely the kind of operation that appeals to their sense of strategy.

Nuclear weapons (NW)

Nuclear weapons are the real threat of mass destruction. Fortunately, the technical barriers to producing enriched uranium or plutonium are prohibitive. Acquiring the technical skill to produce a nuclear warhead is even more difficult, and if successful, one must then have a delivery system, or the ability to transport a radioactive device without detection by sophisticated sensors.

Few states have been willing and able to devote the financial resources to a long-term nuclear weapons program and it would be far beyond the reach of non-state actors.

Nonetheless, it is conceivable that non-state organizations could acquire nuclear weapons by stealing or buying them from a desperate nuclear power, or by seizing power in a weak nuclear state like Pakistan. This still doesn't address the problem of getting a warhead to its target.

The apparent solution is to procure a so-called suitcase nuclear bomb. Although there has been much speculation about suitcase-sized nuclear bombs, it has not been demonstrated that such devices actually exist. In 1997, Alexander Lebed, Russia's former National Security Advisor, claimed that the Soviets had developed such weapons, but doubts remain. According to Fox News, "A 'suitcase' bomb is a very compact and portable nuclear weapon and could have the dimensions of 60 × 40 × 20 cm or 24 × 16 × 8 in.

The smallest possible bomb-like object would be a single critical mass of plutonium (or U-233) at maximum density under normal conditions. The Pu-239 weighs 10.5 kg and is 10.1 cm across. It doesn't take much more than a single critical mass to cause significant explosions ranging from 10 to 20 tons. These types of weapons can also be as big as two footlockers. The warhead consists of a tube with two pieces of uranium, which, when rammed together, would cause a blast. Some sort of firing unit and a device that would need to be decoded to cause detonation may be included in the suitcase."

Although the Russian government has steadfastly denied the existence of man-portable nukes, Russian scientists have testified that they are "absolutely sure" the devices were created. There are also allegations that Osama bin Laden purchased nuclear suitcase and backpack bombs from Chechen organized crime groups, but FBI officials claim there is no evidence to verify the allegations.

If there are no suitcase nukes, then the prospects of them falling into the wrong hands is a moot point. If the Russians did produce such bombs, one can only hope that the Russians have them secured. They do have more than ample motivation. If man-portable nuclear bombs fell into the hands of Islamist terrorists, Russia itself is a likely first target, owing to their policies in Afghanistan and Chechnya.

Given the radiation emitted by a lightly shielded mini-nuke and the existence of highly sensitive sensors, it seems unlikely that they could be smuggled easily.

One of the most feasible conduits, however, would be via shipping containers, the vast majority of which are not inspected.

Given that even a small nuclear device has enormous destructive potential and could be more feasible than large-scale chemical or biological attacks, port security appears to be one area where homeland defense expenditures would be prudent.

While there may be controversy about the definition of the politically-charged word "terrorism," the tactics and technology of chemical terrorism are clearly distinguished from those of other forms of chemical warfare, using chemical weapons designed to meet military needs. Chemical terrorism is asymmetric warfare, as practiced by non-uniformed forces using light and/or improvised weapons against non-combatant targets. It is therefore unlike the symmetric chemical warfare of the First World War, in which dug-in troops fired poison-filled artillery shells at each other across a wire-bounded no-man's-land. It is also distinct from asymmetric "terror from above," in which military forces use munitions with chemical payloads against civilian populations.

Industrial chemicals as weapons of mass destruction

Chemical terrorism is also qualitatively different from biological terrorism involving infectious diseases, but quite similar to the covert employment of biologically-produced toxins, which differ from synthetic poisons mainly in their extreme potency and the means by which they are produced.

There have been few documented acts of chemical terrorism, and none of those has caused casualties justifying the treatment of chemical weapons as weapons of mass destruction.

However, there has been much discussion and some serious study of the possibility of chemical terrorism. One of the stated concerns leading to the 2003 invasion of Iraq was the possibility that chemical weapons technology developed and used by Iraq could be transferred to terrorist organizations.

Tactics

The main issue in chemical warfare, for high-tech, state-funded military users as well as for non-traditional forces, is distributing the material efficiently in the target area. In most chemical warfare scenarios, much or most of the toxic agent will be destroyed by explosive dispersal devices, delivered in massive overkill quantities to a few victims, and/or broadcast into areas where no potential victims exist. If these dispersal devices can be discovered, explosive ordnance disposal teams may render safe the explosive system, or, if that is impossible, conduct a controlled detonation that limits spread of the toxic materials. Toxic agents that do not find victims immediately on delivery may degrade spontaneously, or be deactivated or sequestered by decontamination teams.

No known chemical weapons qualify as "weapons of mass destruction" in the sense of even the Hiroshima and Nagasaki bombs. Realistic chemical attacks will be on a smaller scale, but a campaign of such attacks could be extremely disruptive. The insidious and somewhat mysterious nature of poisons makes them potential weapons of mass terror, because people in a target area—or simply in what they perceive to be a target area—will not know whether or not they've been poisoned.

Methods used by terrorists or hypothesized by analysts include

- Contamination of reservoirs and urban water supply systems.
- Contamination of food, beverages, drugs, or cosmetics in manufacturing or distribution processes.
- Contamination of food or beverages near the point of consumption.
- Miscellaneous product contaminations: stamps/envelopes, IV fluids, and so on.
- Release of gases or aerosols into building HVAC systems.
- Release of gases or aerosols from aircraft.
- Dispersal in bombs or projectiles.
- Miscellaneous direct methods: hand sprayers, water guns, parcels.
- Release of industrial/agricultural chemicals via attacks on production or storage facilities.
- Release of industrial/agricultural chemicals via attacks on truck, rail, or barge shipping.
- Miscellaneous releases of industrial/agricultural chemicals, especially anhydrous ammonia, fumigants and pesticides, and disinfectant gases (e.g., chlorine, chlorine dioxide, ethylene oxide).

Industrial chemicals

Some pesticides share characteristics with military chemical agents. Tabun was originally developed in a German search for new pesticides, and Amiton, the earliest V-series agent, was actually brought to market as a pesticide by a British company. Manufacturing technology and raw materials that can be used to make pesticides may also be used to make chemical warfare agents. The most dangerous pesticides have been largely or completely replaced by more selective alternatives that kill pests effectively with less danger to humans. Several highly toxic "restricted-use" pesticides are still produced and used in very large quantities, and could be hijacked by the truckload. This includes parathion, methyl parathion, and other organ phosphorus compounds with LD50s in the order of

10 mg/kg. Military chemical sensors test for such materials, as well as specific chemical weapons.

As alternatives to highly toxic pesticides become available, the older and more dangerous substances are de-certified for commercial use, and are either dropped entirely or made only for export by the chemical industry. However, some of these are relatively simple compounds that could be made in clandestine labs. For example, TEPP, the first and most dangerous organophosphorus pesticide, though significantly less toxic than tabun, sarin, and VX, is nevertheless fast-acting and deadly enough for use in direct attacks on soft targets, and its relatively simple synthesis is described in old patents.

It is to be expected that certain rodenticides would be extremely effective as contaminants, since their normal application requires them to be stable, odorless, and tasteless, while possessing high mammalian toxicity.

Modern rat and mouse killers meet these criteria without creating extreme hazards for humans. Arsenic, on the other hand, is the classic example of a rat poison that is equally applicable to homicide. Inorganic thallium, barium, and phosphorus compounds might also be used, although some of these will require high concentrations for reliable lethality.

Much more potent than arsenic are two widely banned rodenticides: sodium fluoroacetate, also known as Compound 1080; and tetramethylenedisulfotetramine, sometimes called TETS or "tetramine." The human LD50 for these substances is in the order of 1 mg/kg, with TETS being perhaps 3–10 times stronger than 1080. No antidote is known for either agent. TETS has been widely discussed as a potential terrorist weapon, having been used in several multi-fatality crimes of private revenge in China, where there exists a black market for illegal TETS-based rat poisons made in secret factories. Very small amounts of Compound 1080 are used legally for predator control in the United States, but several tons per year are made for export by Tull Chemical in Alabama. Because the use of 1080 to kill predators has been very controversial, environmental organizations have stressed its potential as a terrorist weapon in their attempts to have it outlawed completely.

Industrial chemicals that can be weaponized

The most common toxic hazardous materials are chlorine and anhydrous ammonia. While chlorine is normally stored and shipped in very large containers, the use of ammonia in agriculture requires it to be distributed to many more sites in smaller containers. These gases create possibilities for highly disruptive large-scale releases, but will cause few or no fatalities unless victims are trapped in areas where concentrations are high.

Recent suicide bombings in Iraq have combined conventional explosives with chlorine tanks; while this technique hasn't produced exceptionally high death tolls, it causes increased chaos and can be expected to drive all unprotected survivors away from the scene of an attack.

Chemical agents and terrorism

The effective use of chemical agents by terrorists is a growing concern among many analysts who believe that it is only a matter of time. The media reports the significant attention now being given to the problems of nuclear proliferation; it is only logical that chemical weapons, as well as biological weapons, should receive the same level of national concern. Although inter-agency and multi-national cooperation in these areas is highly desirable, the actual process of cooperation moves painfully slowly.

Current techniques of detection and decontamination both need to be addressed. Given today's limited resources, decisions need to be made as to where the U.S. government should focus its resources (e.g., intelligence gathering focused on potential proliferates, threat assessment, emergency response capability and the response assets required, countermeasures, federal emergency management procedures, etc.).

Most previous cooperative efforts have historically been DOD-led and focused on the battlefield. The remainder, to include those funded by DHS, have been largely disjointed or short-ranged. That approach is not acceptable and must not be allowed to continue to dominate effective cooperation in the areas of planning, and research and development with regard to chemical weapons identification, detection, and countermeasures for the U.S. homeland.

World War I demonstrated what happened to the soldiers of nations unprepared for chemical warfare. Can anyone contemplate this happening to American soccer moms in a terrorist attack using chemical weapons in a shopping mall? The potential that it may unarguably exists. Now imagine any nation; are we really prepared to handle any kind of large-scale use of any type of chemical agent either military or industrial? As demonstrated by World War I, total devastation may be wreaked on the safety and security of any community by the use of chemical weapons.

Indicators of a possible chemical incident

Dead animals/birds/fish—not just an occasional incident, but numerous animals (wild and domestic, small and large), birds, and fish in the same area

Lack of insect life—normal insect activity (ground, air, and/or water) missing, dead insects evident in the ground/water surface/shoreline

Physical symptoms—numerous individuals experiencing unexplained water-like blisters, wheals (similar to bee stings), pinpointed pupils, choking, respiratory ailments, and/or rashes
Mass casualties—numerous individuals exhibiting unexplained serious health problems ranging from nausea to disorientation, difficulty in breathing, convulsions, and death
Definite pattern of casualties—distributed in a pattern that may be associated with possible agent dissemination methods
Illness associated with confined geographic area—lower incidence of symptoms for people working indoors than outdoors, or the reverse
Unusual liquid droplets—numerous surfaces exhibiting oily droplets/film; numerous water surfaces displaying an oily film (no recent rain)
Areas that look different in appearance—not just a patch of dead weeds, but trees, shrubs, bushes, food crops, and/or lawns that are dead, discolored, or withered (no current drought)
Unexplained odors—smells ranging from fruit/flower to sharp/pungent to garlic/horseradish-like to bitter almonds/peach kernels to newly mown hay; the particular odor is completely out of character with its surroundings
Low-lying clouds—low-lying cloud/fog-like condition that is not explained by its surroundings
Unusual metal debris—unexplained bomb/munitions-like material, especially if it contains a liquid (no recent rain)

Conclusion

The future of WMD entails challenges beyond projecting future developments, because WMD have so seldom been employed, and countries that possess such weapons tend to be secretive about these capabilities and their intentions regarding them. There is a paucity of factual information on why countries will seek WMD and how they are likely to use them. Perhaps the most important and challenging question today is why these weapons have so seldom been used, particularly those chemical, biological, and radiological weapons that are considered accessible to the violent extremists who are believed to be the most likely to employ them.

chapter eight

Explosives

An *explosive material*, also called an explosive, is a reactive substance that contains a great amount of potential energy that can produce an explosion if released suddenly, usually accompanied by the production of light, heat, sound, and pressure. An explosive charge is a measured quantity of explosive material.

This potential energy stored in an explosive material may be

- Chemical energy, such as nitroglycerin or grain dust
- Pressurized gas, such as a gas cylinder or aerosol can.
- Nuclear energy, such as in the fissile isotopes uranium-235 and plutonium-239

Explosive materials may be categorized by the speed at which they expand. Materials that detonate (the front of the chemical reaction moves faster through the material than the speed of sound) are said to be "high explosives," and materials that deflagrate are said to be "low explosives." Explosives may also be categorized by their sensitivity. Sensitive materials that can be initiated by a relatively small amount of heat or pressure are primary explosives, and materials that are relatively insensitive are secondary or tertiary explosives.

A wide variety of chemicals can explode; a smaller number are manufactured in quantity as explosives. The remainder are too dangerous, sensitive, toxic, expensive, unstable, or decompose too quickly for common usage.

Though early thermal weapons, such as Greek fire, have existed since ancient times, the first widely used explosive in warfare and mining was black powder, invented in ninth-century China (see the history of gunpowder). This material was sensitive to water, and it produced dark smoke. The first useful explosive stronger than black powder was nitroglycerin, developed in 1847. As nitroglycerin was unstable, it was replaced by nitrocellulose, smokeless powder, dynamite, and gelignite (the two latter invented by Alfred Nobel). World War I saw the introduction of trinitrotoluene in naval shells. World War II saw an extensive use of new explosives (see explosives used during World War II). In turn, these have largely been replaced by modern explosives such as C-4. The increased availability of chemicals has allowed the construction of improvised explosive devices.

Chemical

An explosion is a type of spontaneous chemical reaction that, once initiated, is driven by both a large exothermic change (great release of heat) and a large positive entropy change (great quantities of gases are released) in going from reactants to products, thereby constituting a thermodynamically favorable process, in addition to one that propagates very rapidly. Thus, explosives are substances that contain a large amount of energy stored in chemical bonds. The energetic stability of the gaseous products, and hence their generation, comes from the formation of strongly bonded species like carbon monoxide, carbon dioxide, and (di)nitrogen, which contain strong double and triple bonds having bond strengths of nearly 1 MJ/mole. Consequently, most commercial explosives are organic compounds containing $-NO_2$, $-ONO_2$, and $-NHNO_2$ groups that, when detonated, release gases like the aforementioned (e.g., nitroglycerin, TNT, HMX, PETN, nitrocellulose).

Decomposition

The chemical decomposition of an explosive may take years, days, hours, or a fraction of a second. The slower processes of decomposition take place in storage and are of interest only from a stability standpoint. Of more interest are the two rapid forms of decomposition, deflagration and detonation.

Deflagration

In deflagration, the decomposition of the explosive material is propagated by a flame front which moves slowly through the explosive material, in contrast to detonation. Deflagration is a characteristic of low-explosive material.

Detonation

This term is used to describe an explosive phenomenon, whereby the decomposition is propagated by an explosive shock wave traversing the explosive material. The shock front is capable of passing through the high-explosive material at great speeds, typically thousands of meters per second. It is any device or material in which the explosive is ignited thereby reaching a reactive incident.

Exotic

In addition to chemical explosives, there are a number of more exotic explosive materials, and exotic methods of causing explosions. Examples include nuclear explosives, and abruptly heating a substance to a plasma state with a high-intensity laser or electric arc.

Laser- and arc-heating are used in laser detonators, exploding-bridge wire detonators, and exploding foil initiators, where a shock wave and then detonation in conventional chemical explosive material is created by laser- or electric-arc heating. Laser and electric energy are not currently used in practice to generate most of the required energy, but only to initiate reactions.

Properties of explosive materials

To determine the suitability of an explosive substance for a particular use, its physical properties must first be known. The usefulness of an explosive can only be appreciated when the properties and the factors affecting them are fully understood. Some of the more important characteristics are listed below.

Availability and cost

The availability and cost of explosives are determined by the availability of the raw materials and the cost, complexity, and safety of the manufacturing operations.

Sensitivity

Sensitivity refers to the ease with which an explosive can be ignited or detonated, that is, the amount and intensity of shock, friction, or heat that is required. When the term sensitivity is used, care must be taken to clarify what kind of sensitivity is under discussion. The relative sensitivity of a given explosive to impact may vary greatly from its sensitivity to friction or heat. Some of the test methods used to determine sensitivity relate to

- *Impact*: Sensitivity is expressed in terms of the distance through which a standard weight must be dropped onto the material to cause it to explode.
- *Friction*: Sensitivity is expressed in terms of what occurs when a weighted pendulum scrapes across the material (it may snap, crackle, ignite, and/or explode).
- *Heat*: Sensitivity is expressed in terms of the temperature at which flashing or explosion of the material occurs.

Sensitivity is an important consideration in selecting an explosive for a particular purpose. The explosive in an armor-piercing projectile must be relatively insensitive, or the shock of impact would cause it to detonate before it penetrated to the point desired. The explosive lenses around nuclear charges are also designed to be highly insensitive, to minimize the risk of accidental detonation.

Sensitivity to initiation

The index of the capacity of an explosive to be detonated is defined by the power of the detonator which is certain to prime the explosive to a sustained and continuous detonation. Reference is made to the Sellier-Bellot scale that consists of a series of 10 detonators, from No. 1 to No. 10, each of which corresponds to an increasing charge weight. In practice, most of the explosives on the market today are sensitive to a No. 8 detonator, where the charge corresponds to 2 g of mercury fulminate.

Velocity of detonation

The velocity with which the reaction process propagates in the mass of the explosive. Most commercial mining explosives have detonation velocities ranging from 1800 to 8000 m/s. Today, velocity of detonation can be measured with accuracy. Together with density, it is an important element influencing the yield of the energy transmitted for both atmospheric overpressure and ground acceleration.

Stability

Stability is the ability of an explosive to be stored without deterioration.

The following factors affect the stability of an explosive:

- *Chemical constitution*. In the strictest technical sense, the word "stability" is a thermodynamic term referring to the energy of a substance relative to a reference state or to some other substance. However, in the context of explosives, stability commonly refers to ease of detonation, which is concerned with kinetics (i.e., rate of decomposition).
- It is perhaps best, then, to differentiate between the terms thermodynamically stable and kinetically stable by referring to the former as "inert." Contrarily, a kinetically unstable substance is said to be "labile." It is generally recognized that certain groups, like nitro ($-NO_2$), nitrate ($-ONO_2$), and azide ($-N_3$), are intrinsically labile. Kinetically, there exists a low activation barrier to the decomposition reaction.
- Consequently, these compounds exhibit high sensitivity to flame or mechanical shock. The chemical bonding in these compounds is characterized as predominantly covalent and thus they are not thermodynamically stabilized by high ionic-lattice energy. Furthermore,

they generally have positive enthalpies of formation and there is little mechanistic hindrance to internal molecular rearrangement to yield the more thermodynamically stable (more strongly bonded) decomposition products; for example, in lead azide, $Pb(N_3)_2$, the nitrogen atoms are already bonded to one another, so decomposition into Pb and N_2. is relatively easy.

- *Temperature of storage*. The rate of decomposition of explosives increases at higher temperatures. All standard military explosives may be considered to have a high degree of stability at temperatures from −10°C to +35°C, but each has a high temperature at which its rate of decomposition rapidly accelerates and stability is reduced. As a rule of thumb, most explosives become dangerously unstable at temperatures above 70°C.
- *Exposure to sunlight*. When exposed to the ultraviolet rays of sunlight, many explosive compounds containing nitrogen groups rapidly decompose, affecting their stability.
- *Electrical discharge*. Electrostatic or spark sensitivity to initiation is common in a number of explosives. Static or other electrical discharge may be sufficient to cause a reaction, even detonation, under some circumstances. As a result, safe handling of explosives and pyrotechnics usually requires proper electrical grounding of the operator.

Power, performance, and strength

The term *power* or *performance* as applied to an explosive refers to its ability to do work. In practice, it is defined as the explosive's ability to accomplish what is intended in the way of energy delivery (i.e., fragment projection, air blast, high-velocity jet, underwater shock and bubble energy, etc.). Explosive power or performance is evaluated by a tailored series of tests to assess the material for its intended use. Of the tests listed below, cylinder expansion and air-blast tests are common to most testing programs, and the others support specific applications.

- *Cylinder expansion test*. A standard amount of explosive is loaded into a long hollow cylinder, usually of copper, and detonated at one end. Data is collected concerning the rate of radial expansion of the cylinder and the maximum cylinder wall velocity. This also establishes the Gurney energy, or $2E$.
- *Cylinder fragmentation*. A standard steel cylinder is loaded with explosive and detonated in a sawdust pit. The fragments are collected and the size distribution analyzed.
- *Detonation pressure (Chapman-Jouguet condition)*. Detonation pressure data derived from measurements of shock waves transmitted into

water by the detonation of cylindrical explosive charges of a standard size.

- *Determination of critical diameter*. This test establishes the minimum physical size a charge of a specific explosive must be to sustain its own detonation wave. The procedure involves the detonation of a series of charges of different diameters until difficulty in detonation wave propagation is observed.
- *Infinite-diameter detonation velocity*. Detonation velocity is dependent on loading density (c), charge diameter, and grain size. The hydrodynamic theory of detonation used in predicting explosive phenomena does not include the diameter of the charge, and therefore a detonation velocity, for an imaginary charge of infinite diameter. This procedure requires the firing of a series of charges of the same density and physical structure, but different diameters, and the extrapolation of the resulting detonation velocities to predict the detonation velocity of a charge of infinite diameter.
- *Pressure versus scaled distance*. A charge of a specific size is detonated and its pressure effects measured at a standard distance. The values obtained are compared with those for TNT.
- *Impulse versus scaled distance*. A charge of a specific size is detonated and its impulse (the area under the pressure-time curve) measured as a function of distance. The results are tabulated and expressed as TNT equivalents.
- *Relative bubble energy (RBE)*. A 5–50 kg charge is detonated in water and piezoelectric gauges measure peak pressure, time constant, impulse, and energy.

The RBE may be defined as K_x 3

$$\text{RBE} = K_s$$

where K = the bubble expansion period for an experimental (x) or a standard (s) charge.

Brisance

In addition to strength, explosives display a second characteristic, which is their shattering effect, or brisance (from the French meaning to "break"), which is distinguished and separate from their total work capacity. This characteristic is of practical importance in determining the effectiveness of an explosion in fragmenting shells, bomb casings, grenades, and the like. The rapidity with which an explosive reaches its peak pressure (power) is a measure of its brisance. Brisance values are primarily employed in France and Russia.

The sand crush test is commonly employed to determine the relative brisance in comparison to TNT. No test is capable of directly comparing the explosive properties of two or more compounds; it is important to examine the data from several such tests (sand crush, Trauzl, and so forth) in order to gauge relative brisance. True values for comparison require field experiments.

Density

Density of loading refers to the mass of an explosive per unit volume. Several methods of loading are available, including pellet loading, cast loading, and press loading, the choice being determined by the characteristics of the explosive. Dependent upon the method employed, an average density of the loaded charge can be obtained that is within 80%–99% of the theoretical maximum density of the explosive.

High load density can reduce sensitivity by making the mass more resistant to internal friction. However, if density is increased to the extent that individual crystals are crushed, the explosive may become more sensitive. Increased load density also permits the use of more explosive, thereby increasing the power of the warhead. It is possible to compress an explosive beyond a point of sensitivity, known also as *dead-pressing*, in which the material is no longer capable of being reliably initiated, if at all.

Volatility

Volatility is the readiness with which a substance vaporizes. Excessive volatility often results in the development of pressure within rounds of ammunition and separation of mixtures into their constituents. Volatility affects the chemical composition of the explosive such that a marked reduction in stability may occur, which results in an increase in the danger of handling.

Hygroscopicity and water resistance

The introduction of water into an explosive is highly undesirable, since it reduces the sensitivity, strength, and velocity of detonation of the explosive. Hygroscopicity is used as a measure of a material's moisture-absorbing tendencies. Moisture affects explosives adversely by acting as an inert material that absorbs heat when vaporized, and by acting as a solvent medium that can cause undesired chemical reactions. Sensitivity, strength, and velocity of detonation are reduced by inert materials that reduce the continuity of the explosive mass. When the moisture content evaporates during detonation, cooling occurs, which reduces the temperature of reaction. Stability is also affected by the presence of moisture, since moisture promotes decomposition of the explosive and, in addition, causes corrosion of the explosive's metal container.

Explosives considerably differ from one another as to their behavior in the presence of water. Gelatin dynamites containing nitroglycerine have a degree of water resistance. Explosives based on ammonium nitrate have little or no water resistance due to the reaction between ammonium nitrate and water, which liberates ammonia, nitrogen dioxide and hydrogen peroxide. In addition, ammonium nitrate is hygroscopic, susceptible to damp, hence the above concerns.

Toxicity

There are many types of explosives which are toxic to some extent. Manufacturing inputs can also be organic compounds or hazardous materials that require special handing due to risks (such as carcinogens). The decomposition products, residual solids or gases of some explosives can be toxic, whereas others are harmless, such as carbon dioxide and water. Examples of harmful by-products are

- Heavy metals, such as lead, mercury and barium from primers (observed in high-volume firing ranges)
- Nitric oxides from TNT
- Per chlorates when used in large quantities

Green explosives seek to reduce environment and health impacts. An example of such is the lead-free primary explosive copper (I) 5-nitrotetrazolate, an alternative to lead azide.

Explosive train

Explosive material may be incorporated in the explosive train of a device or system. An example is a pyrotechnic lead igniting a booster, which causes the main charge to detonate.

Volume of products of explosion

The most widely used explosives are condensed liquids or solids converted to gaseous products by explosive chemical reactions and the energy released by those reactions. The gaseous products of complete reaction are typically carbon dioxide, steam, and nitrogen. Gaseous volumes computed by the ideal gas law tend to be too large at high pressures characteristic of explosions. Ultimate volume expansion may be estimated at three orders of magnitude, or 1 l/g of explosive. Explosives with an oxygen deficit will generate soot or gases like carbon monoxide and hydrogen, which may react with surrounding materials such as atmospheric oxygen. Attempts to obtain more precise volume estimates must consider the possibility of such side reactions, condensation of steam, and aqueous solubility of gases like carbon dioxide.

Oxygen balance (OB% or Ω)

Oxygen balance is an expression that is used to indicate the degree to which an explosive can be oxidized. If an explosive molecule contains just enough oxygen to convert its entire carbon to carbon dioxide, all of its hydrogen to water, and all of its metal to metal oxide with no excess, the molecule is said to have a zero oxygen balance. The molecule is said to have a positive oxygen balance if it contains more oxygen than is needed, and a negative oxygen balance if it contains less oxygen than is needed.

The sensitivity, strength, and brisance of an explosive are all somewhat dependent upon oxygen balance and tend to approach their maxima as oxygen balance approaches zero.

Chemical composition

A chemical explosive may consist of either a chemically pure compound, such as nitroglycerin, or a mixture of a fuel and an oxidizer, such as black powder or grain dust and air.

Chemically pure compounds

Some chemical compounds are unstable in that, when shocked, they react, possibly to the point of detonation. Each molecule of the compound dissociates into two or more new molecules (generally gases) with the release of energy.

- *Nitroglycerin*: A highly unstable and sensitive liquid
- *Acetone peroxide*: A very unstable white organic peroxide
- *TNT*: Yellow insensitive crystals that can be melted and cast without detonation
- *Nitrocellulose*: A nitrated polymer which can be a high or low explosive depending on nitration level and conditions
- *RDX, PETN, HMX*: Very powerful explosives which can be used pure or in plastic explosives
    - *C-4* (or Composition C-4): An RDX plastic explosive plasticized to be adhesive and malleable

The above compositions may describe most of the explosive material, but a practical explosive will often include small percentages of other substances. For example, dynamite is a mixture of highly sensitive nitroglycerin with sawdust, powdered silica, or most commonly, diatomaceous earth, which act as stabilizers. Plastics and polymers may be added to bind powders of explosive compounds; waxes may be incorporated to make them safer to handle; aluminum powder may be introduced to increase total energy and blast effects. Explosive compounds are also often "alloyed": HMX or RDX powders may be mixed (typically by melt-casting) with TNT to form Octol or Cyclotol.

Mixture of oxidizer and fuel

An oxidizer is a pure substance (molecule) that, in a chemical reaction, can contribute some atoms of one or more oxidizing elements, in which the fuel component of the explosive burns. On the simplest level, the oxidizer may itself be an oxidizing element, such as gaseous or liquid oxygen.

- *Black powder*: Potassium nitrate, charcoal, and sulfur
- *Flash powder*: Fine metal powder (usually aluminum or magnesium) and a strong oxidizer (e.g., potassium chlorate or per chlorate)
- *Ammonal*: Ammonium nitrate and aluminum powder
- *Armstrong's mixture*: Potassium chlorate and red phosphorus. This is a very sensitive mixture. It is a primary high explosive in which sulfur is substituted for some or all of the phosphorus to slightly decrease sensitivity.
- *Sprengel explosives*: A very general class incorporating any strong oxidizer and highly reactive fuel, although in practice the name was most commonly applied to mixtures of chlorates and nitro aromatics.
    - *ANFO*: Ammonium nitrate and fuel oil
    - *Cheddites*: Chlorates or per chlorates and oil
    - *Oxyliquits*: Mixtures of organic materials and liquid oxygen
    - *Panclastites*: Mixtures of organic materials and dinitrogen tetroxide

Classification of explosive materials

By sensitivity

Primary explosive

A *primary explosive* is an explosive that is extremely sensitive to stimuli such as impact, friction, heat, static electricity, or electromagnetic radiation. A relatively small amount of energy is required for initiation. As a very general rule, primary explosives are considered to be those compounds that are more sensitive than PETN. As a practical measure, primary explosives are sufficiently sensitive that they can be reliably initiated with a blow from a hammer; however, PETN can also usually be initiated in this manner, so this is only a very broad guideline. Additionally, several compounds, such as nitrogen triiodide, are so sensitive that they cannot even be handled without detonating. Nitrogen triiodide is so sensitive that it can be reliably detonated by exposure to alpha radiation; it is the only explosive for which this is true.

Primary explosives are often used in detonators or to trigger larger charges of less sensitive secondary explosives. Primary explosives are commonly used in blasting caps and percussion caps to translate a physical shock signal. In other situations, different signals such as electrical/physical shock, or in the case of laser detonation systems, light, are used to initiate an action, that is, an explosion. A small quantity, usually milligrams, is sufficient to initiate a larger charge of explosive that is usually safer to handle (Table 8.1).

Secondary explosive

A *secondary explosive* is less sensitive than a primary explosive and requires substantially more energy to be initiated. Because they are less sensitive, they are usable in a wider variety of applications, and are safer to handle and store. Secondary explosives are used in larger quantities in an explosive train and are usually initiated by a smaller quantity of a primary explosive.

Examples of secondary explosives include TNT and RDX.

Tertiary explosive

Tertiary explosives, also called *blasting agents*, are so insensitive to shock that they cannot be reliably detonated by practical quantities of primary explosive, and instead require an intermediate explosive booster of secondary explosive. These are often used for safety and the typically lower costs of material and handling. Primary users are large-scale mining and construction operations. They have also been used for terrorist attacks, because of the sometimes ready availability of large quantities of precursors (e.g., nitrate fertilizers).

ANFO is an example of a tertiary explosive.

By velocity

Low explosives

Low explosives are compounds where the rate of decomposition proceeds through the material at less than the speed of sound. The decomposition is propagated by a flame front (deflagration) which travels much

Table 8.1 Examples of primary high explosives are

- Acetone peroxide
- Alkali metal ozonides
- Ammonium permanganate
- Ammonium chlorate
- Azidotetrazolates
- Azo-clathrates
- Benzoyl peroxide
- Benzvalene
- Chlorine azide
- Chlorine oxides
- Copper(I) acetylide
- Copper(II) azide
- Cumene hydroperoxide
- Cyanogen azide
- Diacetyl peroxide
- Diazodinitrophenol
- Diazomethane
- Diethyl ether peroxide
- 4-Dimethylaminophenylpentazole
- Disulfur dinitride
- Ethyl azide
- Explosive antimony
- Fluorine azide
- Fluorine perchlorate
- Fulminic acid
- Hexamethylene triperoxide diamine
- Hydrazoic acid
- Hypofluorous acid
- Lead azide
- Lead styphnate
- Lead picrate
- Manganese heptoxide
- Mercury(II) fulminate
- Mercury nitride
- Methyl ethyl ketone peroxide
- Nitrogen trichloride
- Nitrogen tribromide
- Nitrogen triiodide
- Nitroglycerin
- Nitronium perchlorate

(*Continued*)

Table 8.1 (Continued) Examples of primary high explosives are

• Nitrotetrazolate-N-oxides
• Octaazacubane
• Pentazenium hexafluoroarsenate
• Peroxy acids
• Peroxymonosulfuric acid
• Selenium tetraazide
• Silicon tetraazide
• Silver azide
• Silver acetylide
• Silver fulminate
• Silver nitride
• Sodium azide
• Tellurium tetraazide
• *tert*-Butyl hydroperoxide
• Tetraamine copper complexes
• Tetraazidomethane
• Tetrazene explosive
• Tetranitratoxycarbon
• Tetrazoles
• Titanium tetraazide
• Triazidomethane
• Xenon dioxide
• Xenon oxytetrafluoride
• Xenon tetroxide
• Xenon trioxide

more slowly through the explosive material than a shock wave of a high explosive. Under normal conditions, low explosives undergo deflagration at rates that vary from a few centimeters per second to approximately 400 m/s.

It is possible for them to deflagrate very quickly, producing an effect similar to a detonation. This can happen under higher pressure or temperature, which usually occurs when ignited in a confined space.

A low explosive is usually a mixture of a combustible substance and an oxidant that decomposes rapidly (deflagration); however, they burn more slowly than a high explosive, which has an extremely fast burn rate.

Low explosives are normally employed as propellants. Included in this group are gunpowders and light pyrotechnics, such as flares and fireworks, but they can replace high explosives in certain applications, see gas pressure blasting.

High explosives

High explosives are explosive materials that detonate, meaning that the explosive shock front passes through the material at a supersonic speed. High explosives detonate with explosive velocity ranging from 3 to 9 km/s. They are normally employed in mining, demolition, and military applications. They can be divided into two explosives classes differentiated by sensitivity: primary explosive and secondary explosive. The term *high explosive* is in contrast with the term *low explosive*, which explodes (deflagrates) at a lower rate.

By composition

Priming composition

Priming compositions are primary explosives mixed with other compositions to control (lessen) the sensitivity of the mixture to the desired level.

For example, primary explosives are so sensitive that they need to be stored and shipped in a wet state to prevent accidental initiation.

By physical form

Explosives are often characterized by the physical form that the explosives are produced or used in. These use forms are commonly categorized as

- Pressings
- Castings
- Plastic or polymer bonded
- Putties (a.k.a. plastic explosives)
- Rubberized
- Extrudable
- Binary
- Blasting agents
- Slurries and gels
- Dynamites

Shipping label classifications

Shipping labels and tags may include both United Nations and national markings.

United Nations markings include numbered Hazard Class and Division (HC/D) codes and alphabetic Compatibility Group codes. Though the two are related, they are separate and distinct. Any Compatibility Group designator can be assigned to any Hazard Class and Division. An example of this hybrid marking would be a consumer firework, which is labeled as 1.4G or 1.4S.

Examples of national markings would include United States Department of Transportation (U.S. DOT) codes.

United Nations Organization (UNO) Hazard Class and Division (HC/D)

Explosives warning sign

The Hazard Class and Division (HC/D) is a numeric designator within a hazard class indicating the explosive's character, predominance of associated hazards, and potential for causing personnel casualties and property damage. It is an internationally accepted system that communicates using, as the minimum markings, the primary hazard associated with a substance.

Listed below are the Divisions for Class 1 (Explosives):

- *1.1* Mass detonation hazard. With HC/D 1.1, it is expected that if one item in a container or pallet inadvertently detonates, the explosion will sympathetically detonate the surrounding items. The explosion could propagate to all or the majority of the items stored together, causing a mass detonation. There will also be fragments from the item's casing and/or structures in the blast area.
- *1.2* Non-mass explosion, fragment-producing. HC/D 1.2 is further divided into three subdivisions, HC/D 1.2.1, 1.2.2 and 1.2.3, to account for the magnitude of the effects of an explosion.
- *1.3* Mass fire, minor blast or fragment hazard. Propellants and many pyrotechnic items fall into this category. If one item in a package or stack initiates, it will usually propagate to the other items, creating a mass fire.
- *1.4* Moderate fire, no blast or fragment. HC/D 1.4 items are listed in the table as explosives with no significant hazard. Most small arms and some pyrotechnic items fall into this category. If the energetic material in these items inadvertently initiates, most of the energy and fragments will be contained within the storage structure or the item containers themselves.
- *1.5* mass detonation hazards, very insensitive.
- *1.6* detonation hazard without mass detonation hazard, extremely insensitive.

Class 1 Compatibility Group

Compatibility Group codes are used to indicate storage compatibility for HC/D Class 1 (explosive) materials. Letters are used to designate 13 compatibility groups as follows.

A. Primary explosive substance (1.1A).
B. An article containing a primary explosive substance and not containing two or more effective protective features. Some articles, such as detonator assemblies for blasting and primers, cap-type, are included. (1.1B, 1.2B, 1.4B).
C. Propellant explosive substance or other deflagrating explosive substance or article containing such explosive substance (1.1C, 1.2C, 1.3C, 1.4C). These are bulk propellants, propelling charges, and devices containing propellants with or without means of ignition. Examples include single-, double-, triple-based, and composite propellants, solid propellant rocket motors, and ammunition with inert projectiles.
D. Secondary detonating explosive substance or black powder or article containing a secondary detonating explosive substance, in each case without means of initiation and without a propelling charge, or article containing a primary explosive substance and containing two or more effective protective features. (1.1D, 1.2D, 1.4D, 1.5D).
E. Article containing a secondary detonating explosive substance without means of initiation, with a propelling charge (other than one containing flammable liquid, gel, or hypergolic liquid) (1.1E, 1.2E, 1.4E).
F. Containing a secondary detonating explosive substance with its means of initiation, with a propelling charge (other than one containing flammable liquid, gel, or hypergolic liquid) or without a propelling charge (1.1F, 1.2F, 1.3F, 1.4F).
G. Pyrotechnic substance or article containing a pyrotechnic substance, or article containing both an explosive substance and an illuminating, incendiary, tear-producing or smoke-producing substance (other than a water-activated article or one containing white phosphorus, phosphide or flammable liquid, gel, or hypergolic liquid) (1.1G, 1.2G, 1.3G, 1.4G). Examples include flares, signals, incendiary or illuminating ammunition, and other smoke and tear-producing devices.
H. Article containing both an explosive substance and white phosphorus (1.2H, 1.3H). These articles will spontaneously combust when exposed to the atmosphere.
J. Article containing both an explosive substance and flammable liquid or gel (1.1J, 1.2J, 1.3J). This excludes liquids or gels which are spontaneously flammable when exposed to water or the atmosphere, which belong in group H. Examples include liquid- or gel-filled incendiary ammunition, fuel-air explosive (FAE) devices, and flammable liquid-fueled missiles.

K. Article containing both an explosive substance and a toxic chemical agent (1.2K, 1.3K)
L. Explosive substance or article containing an explosive substance and presenting a special risk (e.g., due to water-activation or presence of hypergolic liquids, phosphides, or pyrophoric substances) needing isolation of each type (1.1L, 1.2L, 1.3L). Damaged or suspect ammunition of any group belongs in this group.
M. Click here to enter text.
N. Articles containing only extremely insensitive detonating substances (1.6N).
S. Substance or article so packed or designed that any hazardous effects arising from accidental functioning are limited to the extent that they do not significantly hinder or prohibit firefighting or other emergency response efforts in the immediate vicinity of the package (1.4S).

Commercial application

The largest commercial application of explosives is mining. Whether the mine is on the surface, or buried deep underground, there are often times when the use of either a high or low explosive (detonation or deflagration) in a confined space can be used to liberate a fairly specific sub-volume of a brittle material in a much larger volume of the same or similar material. Normally, the material we are talking about in mining is a ceramic of some kind. If there are mineral deposits where large masses of native metal (usually copper) are present in the ground, using explosives to "liberate" the ore typically doesn't work well.

In Materials Science and Engineering, we occasionally see explosives used in cladding. A thin layer of some material is placed on top of a thick layer of a different material, both layers typically being metal. On top of the thin layer is placed an explosive. At one end of the layer of explosive, the explosion is initiated. The two metallic layers are forced together at high speed and with great force. The explosion spreads from the initiation site throughout the entire explosive. Ideally, we hope to produce a metallurgical bond between the two metallic layers.

As the length of time the shock wave spends at any point is small, we can see mixing of the two metals and their surface chemistries, through some fraction of the depth. It is possible that some fraction of the surface material from either layer eventually gets ejected when the end of material is reached. Hence, the mass of the now "welded" bilayer may be less than the sum of the masses of the two initial layers.

There are applications where a shock wave, and electrostatics, can result in high-velocity projectiles.

Explosive velocity

Also known as *detonation velocity* or velocity of detonation (VoD), is the velocity at which the shock wave front travels through a detonated explosive. The data listed for a specific substance is usually a rough prediction based upon gas behavior theory (see Chapman-Jouguet condition), as in practice it is difficult to measure. Explosive velocities are always faster than the local speed of sound in the material.

If the explosive is confined before detonation, such as in an artillery shell, the force produced is focused on a much smaller area, and the pressure is massively intensified. The result is an explosive velocity that is higher than if the explosive had been detonated in the open air: unconfined velocities are often approximately 70%–80% of confined velocities.

Explosive velocity is increased with smaller particle size (i.e., increased spatial density), increased charge diameter, and increased confinement (i.e., higher pressure).

Typical detonation velocities in gases range from 1800 m/s to 3000 m/s. Typical velocities in solid explosives often range beyond 4,000 to 10,300 m/s.

Brisance is the shattering capability of a high explosive, determined mainly by its detonation pressure. The term can be traced from the French verb "*briser*" (to break or shatter), ultimately derived from the Celtic word "*brissim*" (to break). Brisance is of practical importance for determining the effectiveness of an explosion in fragmenting shells, bomb casings, grenades, structures, and the like. The sand crush test and Trauzl lead block test are commonly used to determine the relative brisance in comparison to TNT (which is considered a standard reference for many purposes).

Fragmentation occurs by the action of the transmitted shock wave, the strength of which depends on the detonation pressure of the explosive. Generally, the higher this pressure, the finer the fragments generated. High detonation pressure correlates with high detonation velocity, the speed at which the detonation wave propagates through the explosive, but not necessarily with the explosive's total energy (or work capacity), some of which may be released after passage of the detonation wave. A more brisant explosive, therefore, projects smaller fragments but not necessarily at a higher velocity than a less brisant one.

One of the most brisant of the conventional explosives is cyclotrimethylene trinitramine (also known as RDX or Hexogen). RDX is the explosive agent in the plastic explosive commonly known as C-4, comprising 91% of it by mass.

Detonation involves a supersonic exothermic front accelerating through a medium that eventually drives a shock front propagating directly in front of it. Detonations occur in both conventional solid and liquid explosives, as well as in reactive gases. The velocity of detonations in solid and liquid explosives is much higher than in gaseous ones, which allows the wave system to be observed with greater detail (higher resolution).

An extraordinary variety of fuels may occur as gases, droplet fogs, or dust suspensions. Oxidants include halogens, ozone, hydrogen peroxide, and oxides of nitrogen. Gaseous detonations are often associated with a mixture of fuel and oxidant in a composition somewhat below conventional flammability ratios. They happen most often in confined systems, but they sometimes occur in large vapor clouds.

Explosion is a rapid increase in volume and release of energy in an extreme manner, usually with the generation of high temperatures and the release of gases. Supersonic explosions created by high explosives are known as detonations and travel via supersonic shock waves. Subsonic explosions are created by low explosives through a slower burning process known as deflagration.

Natural

Explosions can occur in nature. Most natural explosions arise from volcanic processes of various sorts. Explosive volcanic eruptions occur when magma rising from below has much dissolved gas in it; the reduction of pressure as the magma rises causes the gas to bubble out of solution, resulting in a rapid increase in volume. Explosions also occur as a result of impact events and in phenomena such as hydrothermal explosions (also due to volcanic processes). Explosions can also occur outside of the Earth, in the universe, in events such as supernovae. Explosions frequently occur during bushfires in eucalyptus forests where the volatile oils in the tree tops suddenly combust.

Animal bodies can also be explosive, as some animals hold a large amount of flammable material such as fat. This, in rare cases, results in naturally exploding animals.

Astronomical

Among the largest known explosions in the universe are supernovae, which result when a star explodes from the sudden starting or stopping of nuclear fusion, and gamma ray bursts, whose nature is still in some dispute. Solar flares are an example of explosions common on the Sun, and presumably on most other stars as well. The energy source for solar flare activity comes from the tangling of magnetic field lines resulting from the rotation of the Sun's conductive plasma. Another type of large astronomical explosion occurs when a very large meteoroid or an asteroid impacts the surface of another object, such as a planet.

Chemical

The most common artificial explosives are chemical explosives, usually involving a rapid and violent oxidation reaction that produces large amounts of hot gas. Gunpowder was the first explosive to be discovered and put to use. Other notable early developments in chemical explosive technology were Frederick Augustus Abel's development of nitrocellulose in 1865 and Alfred Nobel's invention of dynamite in 1866. Chemical explosions (both intentional and accidental) are often initiated by an electric spark or flame. Accidental explosions may occur in fuel tanks or rocket engines.

Electrical and magnetic

A high-current electrical fault can create an "electrical explosion" by forming a high-energy electrical arc which rapidly vaporizes metal and insulation material. This arc flash hazard is a danger to persons working on energized switchgear. Also, excessive magnetic pressure within an ultra-strong electromagnet can cause a *magnetic explosion*.

Mechanical and vapor

Strictly a physical process, as opposed to chemical or nuclear, for example, the bursting of a sealed or partially sealed container under internal pressure is often referred to as a "mechanical explosion." Examples include an overheated boiler or a simple tin can of beans tossed into a fire.

Boiling liquid expanding vapor explosions are one type of mechanical explosion that can occur when a vessel containing a pressurized liquid is ruptured, causing a rapid increase in volume as the liquid evaporates.

Note that the contents of the container may cause a subsequent chemical explosion, the effects of which can be dramatically more serious, such as a propane tank in the midst of a fire. In such a case, to the effects of the mechanical explosion when the tank fails are added the effects from the explosion resulting from the released (initially liquid and then almost instantaneously gaseous) propane in the presence of an ignition source. For this reason, emergency workers often differentiate between the two events.

Nuclear

In addition to stellar nuclear explosions, a man-made nuclear weapon is a type of explosive weapon that derives its destructive force from nuclear fission or from a combination of fission and fusion. As a result, even a nuclear weapon with a small yield is significantly more powerful than the largest conventional explosives available, with a single weapon capable of completely destroying an entire city.

Properties of explosions

Force

Explosive force is released in a direction perpendicular to the surface of the explosive. If the surface is cut or shaped, the explosive forces can be focused to produce a greater local effect; this is known as a shaped charge.

Velocity

The speed of the reaction is what distinguishes the explosive reaction from an ordinary combustion reaction. Unless the reaction occurs rapidly, the thermally expanded gases will be dissipated in the medium, and there will be no explosion. Again, consider a wood or coal fire. As the fire burns, there is the evolution of heat and the formation of gases, but neither is liberated rapidly enough to cause an explosion. This can be likened to the difference between the energy discharge of a battery, which is slow, and that of a flash capacitor like that in a camera flash, which releases its energy all at once.

Evolution of heat

The generation of heat in large quantities accompanies most explosive chemical reactions. The exceptions are called entropic explosives and include organic peroxides, such as acetone peroxide. It is the rapid liberation of heat that causes the gaseous products of most explosive reactions to expand and generate high pressures. This rapid generation of high pressures by the released gas constitutes the explosion.

The liberation of heat with insufficient rapidity will not cause an explosion. For example, although a unit mass of coal yields five times as much heat as a unit mass of nitroglycerin, the coal cannot be used as an explosive (except in the form of coal dust) because the rate at which it yields this heat is quite slow. In fact, a substance which burns less rapidly (slow combustion) may actually evolve more total heat than an explosive which detonates rapidly (i.e., fast combustion). In the former, slow combustion converts more of the internal energy (chemical potential) of the burning substance into heat released to the surroundings, while in the latter, fast combustion (i.e., detonation) instead converts more internal energy into work on the surroundings (i.e., less internal energy converted into heat); *c.f.* heat and work (thermodynamics) are equivalent forms of energy. See *heat of combustion* for a more thorough treatment of this topic.

When a chemical compound is formed from its constituents, heat may either be absorbed or released. The quantity of heat absorbed or given off during transformation is called the heat of formation. Heats of formations for solids and gases found in explosive reactions have been determined for a temperature of 25°C and atmospheric pressure, and are normally given in units of kilojoules per gram-molecule. A negative value indicates that heat is absorbed during the formation of the compound from its elements; such a reaction is called an endothermic reaction. In explosive technology only materials that are exothermic—that have a net liberation of heat—are of interest. Reaction heat is measured under conditions either of constant pressure or constant volume. It is this heat of reaction that may be properly expressed as the "heat of explosion."

Initiation of reaction

A chemical explosive is a compound or mixture which, upon the application of heat or shock, decomposes or rearranges with extreme rapidity, yielding much gas and heat. Many substances not ordinarily classed as explosives may do one, or even two, of these things.

A reaction must be capable of being initiated by the application of shock, heat, or a catalyst (in the case of some explosive chemical reactions) to a small portion of the mass of the explosive material. A material in which the first three factors exist cannot be accepted as an explosive unless the reaction can be made to occur when needed.

Fragmentation

Fragmentation is the accumulation and projection of particles as the result of a high-explosives detonation. Fragments could be part of a structure such as a magazine. High-velocity, low-angle fragments can travel hundreds or thousands of feet with enough energy to initiate other surrounding high-explosive items, injure or kill personnel, and damage vehicles or structures.

Deflagration ("to burn down") is a term describing subsonic combustion propagating through heat transfer; hot burning material heats the next layer of cold material and ignites it. Most "fire" found in daily life, from flames to explosions, is the result of deflagration. Deflagration is different from detonation, which is supersonic, and propagates through shock.

Deflagration is a rapid, high-energy-release combustion event that propagates through a gas or an explosive material at subsonic speeds, driven by the transfer of heat.

Conclusion

Explosives industry members play an integral role in maintaining and improving our quality of life in the United States and work to bring countless benefits to our everyday lives in areas such as mining; oil and gas exploration; demolition; avalanche control; and through the use of explosives in special industrial tools, fire extinguishers, air-bag inflators, fireworks, and special effects in the entertainment industry. However, because of the potential misuse of explosive materials, the ATF plays a vital role in regulating and educating the explosives industry and in protecting the public from inadequate storage and security.

chapter nine

Nuclear weapons and radiation

A brief history of nuclear weapons

A nuclear weapon is an explosive device that derives its destructive force from nuclear reactions, either fission or a combination of fission and fusion. Both reactions release vast quantities of energy from relatively small amounts of matter. The first fission ("atomic") bomb test released the same amount of energy as approximately 20,000 tons of TNT. The first thermonuclear ("hydrogen") bomb test released the same amount of energy as approximately 10,000,000 tons of TNT.

A thermonuclear weapon weighing little more than 2400 lb (1100 kg) can produce an explosive force comparable to the detonation of more than 1.2 million tons (1.1 million tons) of TNT. Thus, even a small nuclear device no larger than traditional bombs can devastate an entire city by blast, fire, and radiation. Nuclear weapons are considered weapons of mass destruction, and their use and control have been a major focus of international relations policy since their debut.

Two nuclear weapons have been used in the course of warfare, both times by the United States near the end of World War II. On August 6, 1945, a uranium gun-type fission bomb, code-named "Little Boy," was detonated over the Japanese city of Hiroshima. Three days later, on August 9, a plutonium implosion-type fission bomb, code-named "Fat Man," was exploded over Nagasaki, Japan. These two bombings resulted in the deaths of approximately 200,000 people—mostly civilians—from acute injuries sustained from the explosions. The role of the bombings in Japan's surrender, and their ethical status, remain the subject of scholarly and popular debate.

Since the bombings of Hiroshima and Nagasaki, nuclear weapons have been detonated on over 2000 occasions for testing purposes and demonstrations. Only a few nations possess such weapons or are suspected of seeking them. The only countries known to have detonated nuclear weapons—and that acknowledge possessing such weapons—are (chronologically by date of first test) the United States, the Soviet Union (succeeded as a nuclear power by Russia), the United Kingdom, France, the People's Republic of China, India, Pakistan, and North Korea. Israel is also widely believed to possess nuclear weapons, though it does not acknowledge having them. One state, South Africa, fabricated nuclear weapons in the past, but as its apartheid regime was coming to an end it disassembled its arsenal, acceded to the Nuclear Non-Proliferation Treaty, and accepted full-scope international safeguards.

The Federation of American Scientists estimates there were more than 17,000 nuclear warheads in the world as of 2012, with around 4,300 of them considered "operational," ready for use.

There are two basic types of nuclear weapons: those that derive the majority of their energy from nuclear fission reactions alone, and those that use fission reactions to begin nuclear fusion reactions that produce a large amount of the total energy output.

Fission weapons

All existing nuclear weapons derive some of their explosive energy from nuclear fission reactions. Weapons whose explosive output is exclusively from fission reactions are commonly referred to as atomic bombs or atom bombs (abbreviated as A-bombs). This has long been noted as something of a misnomer, as their energy comes from the nucleus of the atom, just as it does with fusion weapons.

In fission weapons, a mass of fissile material (enriched uranium or plutonium) is assembled into a supercritical mass—the amount of material needed to start an exponentially growing nuclear chain reaction—either by shooting one piece of sub-critical material into another (the "gun" method) or by compressing a sub-critical sphere of material using chemical explosives to many times its original density (the "implosion" method). The latter approach is considered more sophisticated than the former, and only the latter approach can be used if the fissile material is plutonium.

A major challenge in all nuclear weapon designs is to ensure that a significant fraction of the fuel is consumed before the weapon destroys itself. The amount of energy released by fission bombs can range from the equivalent of just under a ton of TNT to upwards of 500,000 tons (500 kilotons) of TNT.

All fission reactions necessarily generate fission products, the radioactive remains of the atomic nuclei split by the fission reactions. Many fission products are either highly radioactive (but short-lived) or moderately radioactive (but long-lived), and as such are a serious form of radioactive contamination if not fully contained. Fission products are the principal radioactive component of nuclear fallout.

The most commonly used fissile materials for nuclear weapons applications have been uranium-235 and plutonium-239. Less commonly used has been uranium-233. Neptunium-237 and some isotopes of americium may be usable for nuclear explosives as well, but it is not clear that this has ever been implemented, and even their plausible use in nuclear weapons is a matter of scientific dispute.

Fusion weapons

Thermonuclear weapons

The basic principle of the Teller–Ulam design for a hydrogen bomb is the use of radiation derived from a fission bomb to compress and heat a separate section of fusion fuel.

The other basic type of nuclear weapon produces a large proportion of its energy in nuclear fusion reactions. Such fusion weapons are generally referred to as thermonuclear weapons or more colloquially as hydrogen bombs (abbreviated as H-bombs), as they rely on fusion reactions between isotopes of hydrogen (deuterium and tritium). All such weapons derive a significant portion, and sometimes a majority, of their energy from fission. This is because a fission weapon is required as a "trigger" for the fusion reactions, and the fusion reactions can themselves trigger additional fission reactions.

Only six countries—United States, Russia, United Kingdom, People's Republic of China, France and India—have conducted thermonuclear weapon tests. (Whether India has detonated a "true," multi-staged thermonuclear weapon is controversial.) Thermonuclear weapons are considered much more difficult to successfully design and execute than primitive fission weapons. Almost all of the nuclear weapons deployed today use the thermonuclear design because it is more efficient.

Thermonuclear bombs work by using the energy of a fission bomb to compress and heat fusion fuel. In the Teller-Ulam design, which accounts for all multi-megaton yield hydrogen bombs, this is accomplished by placing a fission bomb and fusion fuel (tritium, deuterium, or lithium deuteride) in proximity within a special, radiation-reflecting container. When the fission bomb is detonated, gamma rays and X-rays emitted first compress the fusion fuel, then heat it to thermonuclear temperatures. The ensuing fusion reaction creates enormous numbers of high-speed neutrons, which can then induce fission in materials not normally prone to it, such as depleted uranium. Each of these components is known as a "stage," with the fission bomb as the "primary" and the fusion capsule as the "secondary." In large, megaton-range hydrogen bombs, about half of the yield comes from the final fissioning of depleted uranium.

Virtually all thermonuclear weapons deployed today use the "two-stage" design described above, but it is possible to add additional fusion stages—each stage igniting a larger amount of fusion fuel in the next stage. This technique can be used to construct thermonuclear weapons of arbitrarily large yield, in contrast to fission bombs, which are limited in their explosive force. The largest nuclear weapon ever detonated—the Tsar Bomba of the USSR, which released an energy equivalent of over 50 million tons (50 megatons) of TNT—was a three-stage weapon. Most thermonuclear weapons are considerably smaller than this, due to practical constraints from missile warhead space and weight requirements.

Fusion reactions do not create fission products, and thus contribute far less to the creation of nuclear fallout than fission reactions, but because all thermonuclear weapons contain at least one fission stage, and many high-yield thermonuclear devices have a final fission stage, thermonuclear weapons can generate at least as much nuclear fallout as fission-only weapons.

Nuclear weapons: Other uses

There are other types of nuclear weapons as well. For example, a boosted fission weapon is a fission bomb that increases its explosive yield through a small amount of fusion, but it is not a fusion bomb. In the boosted bomb, the neutrons produced by the fusion reactions serve primarily to increase the efficiency of the fission bomb.

Some weapons are designed for special purposes; a neutron bomb is a thermonuclear weapon that yields a relatively small explosion, but a relatively large amount of neutron radiation; such a device could theoretically be used to cause massive casualties while leaving infrastructure mostly intact and creating a minimal amount of fallout. The detonation of any nuclear weapon is accompanied by a blast of neutron radiation. Surrounding a nuclear weapon with suitable materials (such as cobalt or gold) creates a weapon known as a salted bomb. This device can produce exceptionally large quantities of radioactive contamination.

Research has been done into the possibility of pure fusion bombs: nuclear weapons that consist of fusion reactions without requiring a fission bomb to initiate them. Such a device might provide a simpler path to thermonuclear weapons than one that required development of fission weapons first, and pure fusion weapons would create significantly less nuclear fallout than other thermonuclear weapons, because they would not disperse fission products. In 1998, the United States Department of Energy divulged that the United States had, "made a substantial investment" in the past to develop pure fusion weapons, but that "The U.S. does not have and is not developing a pure fusion weapon," and that "No credible design for a pure fusion weapon resulted from the DoE investment."

Most variation in nuclear weapon design is for the purpose of achieving different yields for different situations, and in manipulating design elements to attempt to minimize weapon size.

Antimatter, which consists of particles resembling ordinary matter particles in most of their properties but having opposite electric charge, has been considered as a trigger mechanism for nuclear weapons. A major obstacle is the difficulty in producing antimatter in large enough quantities, and there is no evidence that it is feasible beyond the military domain. However, the U.S. Air Force funded studies of the physics of antimatter in the Cold War, and began considering its possible use in weapons, not just as a trigger, but as the explosive itself. A fourth-generation nuclear weapon design is related to, and relies upon, the same principle as antimatter-catalyzed nuclear pulse propulsion.

Nuclear weapons delivery systems

Delivering a nuclear weapon to its target is an important aspect of both nuclear weapon design and nuclear strategy. Additionally, development and maintenance of delivery options are among the most resource-intensive aspects of a nuclear weapons program: according to one estimate, deployment costs accounted for 57% of the total financial resources spent by the United States in relation to nuclear weapons since 1940.

Historically the first method of delivery, and the method used in the two nuclear weapons used in warfare, was as a gravity bomb, dropped from bomber aircraft. This is usually the first method that countries develop, as it does not place many restrictions on the size of the weapon: *weapon miniaturization* requires considerable weapons design knowledge. It does, however, limit attack range, response time to an impending attack, and the number of weapons that a country can field at the same time.

With the advent of miniaturization, nuclear bombs can be delivered by both strategic bombers and tactical fighter-bombers, allowing an air force to use its current fleet with little or no modification. This method may still be considered the primary means of nuclear weapons delivery; the majority of U.S. nuclear warheads, for example, are free-fall gravity bombs, namely the B61.

More preferable from a strategic point of view is a nuclear weapon mounted onto a missile, which can use a ballistic trajectory to deliver the warhead over the horizon. Although even short-range missiles allow for a faster and less vulnerable attack, the development of long-range intercontinental ballistic missiles (ICBMs) and submarine-launched ballistic missiles (SLBMs) has given some nations the ability to plausibly deliver missiles anywhere on the globe with a high likelihood of success.

More advanced systems, such as multiple independently targetable reentry vehicles (MIRVs), can launch multiple warheads at different targets from one missile, reducing the chance of a successful missile defense. Today, missiles are most common among systems designed for delivery of nuclear weapons. Making a warhead small enough to fit onto a missile, though, can be difficult.

Tactical weapons have the greatest variety of delivery types, including not only gravity bombs and missiles but also artillery shells, land mines, and nuclear depth charges and torpedoes for anti-submarine warfare. An atomic mortar was also tested at one time by the United States. Small, two-man-portable tactical weapons (somewhat misleadingly referred to as suitcase bombs), such as the Special Atomic Demolition Munition, have been developed, although the difficulty of combining sufficient yield with portability limits their military utility.

Nuclear strategy

Nuclear warfare strategy is a set of policies that deal with preventing or fighting a nuclear war. The policy of trying to prevent an attack by a nuclear weapon from another country by threatening nuclear retaliation is known as the strategy of nuclear deterrence. The goal in deterrence is to always maintain a second-strike capability (the ability of a country to respond to a nuclear attack with one of its own) and potentially to strive for first-strike status (the ability to completely destroy an enemy's nuclear forces before they could retaliate). During the Cold War, policy and military theorists in nuclear-enabled countries worked out models of what sorts of policies could prevent one from ever being attacked by a nuclear weapon, and developed weapon game theory models that create the greatest and most stable deterrence conditions.

The now decommissioned United States' Peacekeeper missile was an ICBM developed to entirely replace the Minuteman missile in the late 1980s. Each missile, like the heavier lift Russian SS-18 Satan, could contain up to 10 nuclear warheads, each of which could be aimed at a different target. A factor in the development of MIRVs was to make complete missile defense very difficult for an enemy country.

Different forms of nuclear weapons delivery allow for different types of nuclear strategies. The goals of any strategy are generally to make it difficult for an enemy to launch a pre-emptive strike against the weapon system and difficult to defend against the delivery of the weapon during a potential conflict. Sometimes this has meant keeping the weapon locations hidden, such as deploying them on submarines or land mobile transporter erector launchers whose locations are very hard for an enemy to track. Alternatively, this can mean protecting them by burying them in hardened missile silo bunkers.

Other components of nuclear strategies have included using missile defense (to destroy the missiles before they land) or implementation of civil defense measures (using early-warning systems to evacuate citizens to safe areas before an attack).

Note that weapons designed to threaten large populations, or to generally deter attacks, are known as strategic weapons. Weapons designed for use on a battlefield in military situations are called tactical weapons.

There are critics of the very idea of strategy for waging nuclear war, who have suggested that a nuclear war between two nuclear powers would result in mutual annihilation. From this point of view, the significance of nuclear weapons is purely to deter war, because any nuclear war would immediately escalate out of mutual distrust and fear, resulting in mutually assured destruction. This threat of national, if not global, destruction has been a strong motivation for anti-nuclear weapons activism.

Critics from the peace movement and within the military establishment have questioned the usefulness of such weapons in the current military climate. According to an advisory opinion issued by the International Court of Justice in 1996, the use of (or threatened use of) such weapons would generally be contrary to the rules of international law applicable in armed conflict, but the court did not reach an opinion as to whether or not the threat or use would be lawful in specific extreme circumstances such as if the survival of the state were at stake.

Another deterrence view in nuclear strategy is that nuclear proliferation can be desirable. This view argues that, unlike conventional weapons, nuclear weapons successfully deter all-out war between states, and they succeeded in doing this during the Cold War between the United States and the Soviet Union. In the late 1950s and early 1960s, Gen. Pierre Marie Gallois of France, an adviser to Charles de Gaulle, argued in books like *The Balance of Terror: Strategy for the Nuclear Age* (1961) that mere possession of a nuclear arsenal, what the French called the *force de frappe*, was enough to ensure deterrence, and thus concluded that the spread of nuclear weapons could increase international stability. Some very prominent neo-realist scholars, such as the late Kenneth Waltz, formerly a Professor of Political Science at UC Berkeley and Adjunct Senior Research Scholar at Columbia University, and John Mearsheimer of the University of Chicago, have also argued along the lines of Gallois.

Specifically, these scholars have advocated some form of nuclear proliferation, arguing that it would decrease the likelihood of total war, especially in troubled regions of the world where there exists a unipolar nuclear weapon state. Aside from the public opinion that opposes proliferation in any form, there are two schools of thought on the matter: those, like Mearsheimer, who favor selective proliferation, and those of Kenneth Waltz, who was somewhat more non-interventionist.

The threat of potentially suicidal terrorists possessing nuclear weapons (a form of nuclear terrorism) complicates the decision process. The prospect of mutually assured destruction may not deter an enemy who expects to die in the confrontation. Further, if the initial act is from a stateless terrorist instead of a sovereign nation, there is no fixed nation or fixed military targets to retaliate against. It has been argued by the New York Times, especially after the September 11, 2001 attacks, that this complication is the sign of the next age of nuclear strategy, distinct from the relative stability of the Cold War. In 1996, the United States adopted a policy of allowing the targeting of its nuclear weapons at terrorists armed with weapons of mass destruction.

Robert Gallucci, president of the John D. and Catherine T. MacArthur Foundation, argues that although traditional deterrence is not an effective approach toward terrorist groups bent on causing a nuclear catastrophe, that "the United States should instead consider a policy of expanded deterrence, which focuses not solely on the would-be nuclear terrorists but on those states that may deliberately transfer or inadvertently lead nuclear weapons and materials to them. By threatening retaliation against those states, the United States may be able to deter that which it cannot physically prevent."

Graham Allison makes a similar case, arguing that the key to expanded deterrence is coming up with ways of tracing nuclear material to the country that forged the fissile material. "After a nuclear bomb detonates, nuclear forensics cops would collect debris samples and send them to a laboratory for radiological analysis. By identifying unique attributes of the fissile material, including its impurities and contaminants, one could trace the path back to its origin." The process is analogous to identifying a criminal by fingerprints. "The goal would be twofold: first, to deter leaders of nuclear states from selling weapons to terrorists by holding them accountable for any use of their own weapons; second, to give leaders every incentive to tightly secure their nuclear weapons and materials."

Nuclear terrorism

Nuclear terrorism denotes the detonation by terrorists of a yield-producing nuclear bomb containing fissile material. Some definitions of nuclear terrorism include the sabotage of a nuclear facility and/or the detonation of a radiological device, colloquially termed a dirty bomb, but consensus is lacking. In legal terms, nuclear terrorism is an offense committed if a person unlawfully and intentionally "uses in any way radioactive material … with the intent to cause death or serious bodily injury; or with the intent to cause substantial damage to property or to the environment; or with the intent to compel a natural or legal person, an international organization or a State to do or refrain from doing an act," according to the 2005 United Nations International Convention for the Suppression of Acts of Nuclear Terrorism.

The possibility of terrorist organizations using nuclear weapons (especially very small ones, such as suitcase nukes) has been a threat in

American rhetoric and culture. It is considered plausible that terrorists could acquire a nuclear weapon. In 2011, the British news agency, the *Telegraph*, received leaked documents regarding the Guantanamo Bay interrogations of Khalid Sheikh Mohammed. The documents cited Khalid saying that, if Osama bin Laden is captured or killed by the Coalition of the Willing, an al-Qaeda sleeper cell would detonate a "weapon of mass destruction" in a "secret location" in Europe, and promised it would be "a nuclear hell storm." This has created much controversy over whether Osama bin Laden was killed, because such a weapon has not been detonated. However, despite some reported thefts and trafficking of small quantities of fissile material, there is no credible evidence that any terrorist group has ever succeeded in obtaining the necessary multi-kilogram critical mass quantities of weapons-grade plutonium required to make a nuclear weapon.

Nuclear terrorism could include

- Acquiring or fabricating a nuclear weapon
- Fabricating a dirty bomb
- Attacking a nuclear reactor, for example, by disrupting critical inputs (e.g., water supply)
- Attacking or taking over a nuclear-armed submarine, plane, or base.

Nuclear terrorism, according to a 2011 report published by the Belfer Center for Science and International Affairs at Harvard University, can be executed and distinguished via four pathways:

- The use of a nuclear weapon that has been stolen or purchased on the black market
- The use of a crude explosive device built by terrorists or by nuclear scientists who the terrorist organization has furtively recruited
- The use of an explosive device constructed by terrorists and their accomplices using their own fissile material
- The acquisition of fissile material from a nation-state.

Militant groups

Nuclear weapons materials on the black market are a global concern, and there is concern about the possible detonation of a small, crude nuclear weapon by a militant group in a major city, with significant loss of life and property.

It is feared that a terrorist group could detonate a radiological or "dirty bomb." A "dirty bomb" is composed of any radioactive source and a conventional explosive. The radioactive material is dispersed by the detonation of the explosive. Detonation of such a weapon is not as powerful as a nuclear blast, but can produce considerable radioactive fallout. There are other radiological weapons called radiological exposure devices, where an explosive is not necessary. A radiological weapon may be very appealing to terrorist groups, as it is highly successful in instilling fear and panic among a population (particularly because of the threat of radiation poisoning), and would contaminate the immediate area for some period of time, disrupting attempts to repair the damage and subsequently inflicting significant economic losses.

According to leaked diplomatic documents, al-Qaeda can produce radiological weapons, after sourcing nuclear material and recruiting rogue scientists to build "dirty bombs." Al-Qaeda, along with some North Caucasus terrorist groups that seek to establish an Islamic Caliphate in Russia, have consistently stated they seek nuclear weapons and have tried to acquire them. Al-Qaeda has sought nuclear weapons for almost two decades by attempting to purchase stolen nuclear material and weapons, and has sought nuclear expertise on numerous occasions. Osama bin Laden has stated that the acquisition of nuclear weapons or other weapons of mass destruction is a "religious duty." While pressure from a wide range of counter-terrorist activity has hampered al-Qaeda's ability to manage such a complex project, there is no sign that it has jettisoned its goals of acquiring fissile material. Statements made as recently as 2008 indicate that al-Qaeda's nuclear ambitions are still very strong.

North Caucasus terrorists have attempted to seize a nuclear submarine armed with nuclear weapons. They have also engaged in reconnaissance activities on nuclear storage facilities and have repeatedly threatened to sabotage nuclear facilities. Similar to al-Qaeda, these groups' activities have been hampered by counter-terrorism activity; nevertheless, they remain committed to launching such a devastating attack within Russia.

The Japanese terror cult Aum Shinrikyo, which used nerve gas to attack a Tokyo subway in 1995, has also tried to acquire nuclear weapons. However, according to nuclear terrorism researchers at Harvard University's Belfer Center for Science and International Affairs, there is no evidence that they continue to do so.

Incidents involving nuclear material

Information reported to the International Atomic Energy Agency (IAEA) shows "a persistent problem with the illicit trafficking in nuclear and other radioactive materials, thefts, losses and other unauthorized activities." The IAEA Illicit Nuclear Trafficking Database notes 1266 incidents reported by 99 countries over the last 12 years, including 18 incidents involving HEU or plutonium trafficking:

- There have been 18 incidences of theft or loss of highly enriched uranium (HEU) and plutonium confirmed by the IAEA.
- Security specialist Shaun Gregory argued in an article that terrorists have attacked Pakistani nuclear facilities three times in the recent past; twice in 2007 and once in 2008.
- In November 2007, burglars with unknown intentions infiltrated the Pelindaba nuclear research facility near Pretoria, South Africa. The burglars escaped without acquiring any of the uranium held at the facility.
- In June 2007, the Federal Bureau of Investigation released to the press the name of Adnan Gulshair el Shukrijumah, allegedly the operations leader for developing tactical plans for detonating nuclear bombs in several American cities simultaneously.
- In November 2006, MI5 warned that al-Qaeda were planning on using nuclear weapons against cities in the United Kingdom by obtaining the bombs via clandestine means.
- In February 2006, Oleg Khinsagov of Russia was arrested in Georgia, along with three Georgian accomplices, with 79.5 g of 89% HEU.
- The Alexander Litvinenko poisoning with radioactive polonium "represents an ominous landmark: the beginning of an era of nuclear terrorism," according to Andrew J. Patterson.
- In June 2002, U.S. citizen José Padilla was arrested for allegedly planning a radiological attack on the city of Chicago; however, he was never charged with such conduct. He was instead convicted of charges that he conspired to "murder, kidnap and maim" people overseas.

Conclusion

If one were to imagine for a moment that commercial nuclear power no longer existed, it would be obvious that the only use a country would then have for its uranium mining, milling, fuel fabrication, and reactors would be to produce nuclear weapons. But because commercial nuclear power do exist, it is sometimes difficult to tell whether a country is using its reactors for research, or for weapons production.

It is precisely this ambiguity which makes the proliferation of nuclear weapons from so-called "peaceful research" a certainty, and the proliferation of commercial nuclear reactors worldwide a Trojan horse for nuclear weapons production.

Since World War II, there have been several instances where countries have pieced together nuclear weapons from the fuel from "peaceful research reactors." France, China, and India have done so. Recently, it was feared that Iraq and North Korea would do likewise, a prospect which was lessened only through the direct threat or actual use of military intervention as an option. Examination of the list of countries currently building or desiring "peaceful" nuclear reactors and the leaders of those nations does not inspire confidence for curtailing nuclear proliferation, either.

It is not just having nuclear weapons which are a threat to peace. In some instances the mere possession or attempted construction of research reactors and commercial nuclear plants has been enough to bring on the threat of war. This "provocation" was enough to justify the Israeli bombing of Iraq's French-built Osirik reactor in 1981, and was one of the alleged reasons for the Gulf War in 1991. The mere inkling that your neighbor might have the capability to make nuclear weapons suddenly becomes the justification for "pre-emptive strikes," and perhaps even full-fledged warfare.

To be sure, there are international agreements and agencies set up to monitor the use of nuclear reactors. The International Atomic Energy Agency is such an entity. However, not all countries have signed agreements allowing inspections by the IAEA. The IAEA itself admitted that even if inspections were allowed, it would not be able to tell if a country was using its commercial reactors to produce weapons.

Table 9.1 Nuclear Weapons by Country

Country	Warheads active/total	Date of first test	CTBT status
The five nuclear-weapon states under the NPT			
United States	2,150/7,700	16 July 1945 ("*Trinity*")	Signatory
Russia	1,800/8,500	29 August 1949 ("*RDS-1*")	Ratified
United Kingdom	160/225	3 October 1952 ("*Hurricane*")	Ratified
France	290/300	13 February 1960 ("*Gerboise Bleue*")	Ratified
China	n.a./250	16 October 1964 ("*596*")	Signatory
Non-NPT nuclear powers			
India	n.a./90–110	18 May 1974 ("*Smiling Buddha*")	Non-signatory
Pakistan	n.a./100–120	28 May 1998 ("*Chagai-I*")	Non-signatory
North Korea	n.a./< 10	9 October 2006[6]	Non-signatory
Undeclared nuclear powers			
Israel	n.a./60–400	Unknown (possibly 22 September 1979)	Signatory

It takes about 15 lb of plutonium-239 or uranium-235 to fashion a crude nuclear device. The technology to enrich the isotopes is available for about one million dollars. It is clearly possible that terrorists could acquire both the isotopes and the technology needed to enrich them. This possibility has surfaced in the news since the breakup of the Soviet Union, and the subsequent revelation of a thriving "black market" in such materials (Table 9.1).

chapter ten

Dirty bomb

A "dirty bomb" is one type of radiological dispersal device (RDD) that combines conventional explosives, such as dynamite, with radioactive material. The terms *dirty bomb* and *RDD* are often used interchangeably in the media. Most RDDs would not release enough radiation to kill people or cause severe illness—the conventional explosive itself would be more harmful to individuals than the radioactive material. However, depending on the situation, an RDD explosion could create fear and panic, contaminate property, and require potentially costly cleanup. Making prompt, accurate information available to the public may prevent the panic sought by terrorists.

A dirty bomb is in no way similar to a nuclear weapon or nuclear bomb. A nuclear bomb creates an explosion that is millions of times more powerful than that of a dirty bomb. The cloud of radiation from a nuclear bomb could spread tens to hundreds of square miles, whereas a dirty bomb's radiation could be dispersed within a few blocks or miles of the explosion. A dirty bomb is not a "weapon of mass destruction" but a "weapon of mass *disruption*," where contamination and anxiety are the terrorists' major objectives.

Impact of a dirty bomb

The extent of local contamination would depend on a number of factors, including the size of the explosive, the amount and type of radioactive material used, the means of dispersal, and weather conditions. Those closest to the RDD would be the most likely to sustain injuries due to the explosion. As radioactive material spreads, it becomes less concentrated and less harmful. Prompt detection of the type of radioactive material used will greatly assist local authorities in advising the community on protective measures, such as sheltering in place, or quickly leaving the immediate area.

Radiation can be readily detected with equipment already carried by many emergency responders. Subsequent decontamination of the affected area may involve considerable time and expense.

Immediate health effects from exposure to the low radiation levels expected from an RDD would likely be minimal. The effects of radiation exposure would be determined by

- The amount of radiation absorbed by the body
- The type of radiation (gamma, beta, or alpha)
- The distance from the radiation to an individual
- The means of exposure—external or internal (absorbed by the skin, inhaled, or ingested); and the duration of exposure

The health effects of radiation tend to be directly proportional to radiation dose. In other words, the higher the radiation dose, the higher the risk of injury.

Protective actions

- In general, protection from radiation is afforded by
- Minimizing the time exposed to radioactive materials
- Maximizing the distance from the source of radiation
- Shielding from external exposure and inhaling radioactive material

Sources of radioactive material

Radioactive materials are routinely used at hospitals, research facilities, industrial activities, and construction sites. These radioactive materials are used for such purposes as diagnosing and treating illnesses, sterilizing equipment, and inspecting welding seams.

Control of radioactive material

Nuclear Regulatory Commission (NRC) and state regulations require owners licensed to use or store radioactive material to secure it from theft and unauthorized access. These measures have been greatly strengthened since the attacks of September 11, 2001. Licensees must promptly report lost or stolen risk-significant radioactive material. "Risk-significant" refers to radioactive sources that may pose a significant risk to individuals, society, and the environment if not properly used, protected, or secured. Local authorities also assist in making a determined effort to find and retrieve such sources. Most reports of lost or stolen material involve small or short-lived radioactive sources that are not useful for an RDD.

Past experience suggests there has not been a pattern of collecting such sources for the purpose of assembling an RDD. It is important to note that, in the United States, the radioactivity of the combined total of all unrecovered sources over the past 8 years (when corrected for radioactive decay) would not reach the threshold for one high-risk radioactive source. Unfortunately, the same cannot be said worldwide.

The U.S. government is working to strengthen security for high-risk radioactive sources both at home and abroad.

The NRC and its 37 Agreement States—states who have been given authority to regulate nuclear materials within their borders—have worked together to create a strong and effective regulatory safety and security framework that includes licensing, inspection, and enforcement.

NRC also works with other federal agencies, the International Atomic Energy Agency, and licensees to protect radioactive material from theft and unauthorized access. The agency has made improvements and upgrades to the joint NRC-DOE (Department of Energy) database that tracks the location and movement of certain forms and quantities of special nuclear material. In addition, in early 2009, NRC deployed its new National Source Tracking System, designed to track high-risk sources in the United States on a continuous basis.

Risk of cancer

Just because a person is near a radioactive source for a short time or gets a small amount of radioactive dust on himself or herself does not mean he or she will get cancer. Any additional risk will likely be extremely small. Doctors specializing in radiation health effects will be able to assess the risks and suggest what medical treatment, if any, is needed, once the radioactive source and exposure levels have been determined.

There are some medical treatments available that help cleanse the body of certain radioactive materials following a radiological accident. Prussian blue has been proven effective for ingestion of cesium-137 (a radioactive isotope). In addition, potassium iodide (KI) can be used to protect against thyroid cancer caused by iodine-131 (radioactive iodine).

However, KI, which is available "over the counter," offers no protection to other parts of the body or against other radioactive isotopes. Medical professionals are best qualified to determine how to best treat symptoms.

Other contact information

A number of federal agencies have responsibilities for dealing with RDDs. Their public affairs offices can answer questions on the subject or provide access to experts in and out of government. Their websites are

- Center for Disease Control and Prevention: www.bt.cdc.gov/radiation
- Department of Homeland Security: www.dhs.gov
- Department of Energy: www.energy.gov/
- Environmental Protection Agency: www.epa.gov
- Nuclear Regulatory Commission: www.nrc.gov

- Federal Emergency Management Agency: www.fema.gov
- Department of Justice: www.usdoj.gov
- Federal Bureau of Investigation: www.fbi.gov
- Department of Health and Human Services: www.hhs.gov
- Transportation Security Administration: www.tsa.gov/public/
- National Nuclear Security Administration: www.nnsa.doe.gov/

What is an RDD or "dirty bomb"?

A "dirty bomb" is one type of a "radiological dispersal device" (RDD) that combines a conventional explosive, such as dynamite, with radioactive material that may disperse when the device explodes. It is not the same as a nuclear weapon. If there are casualties, they will be caused by the initial blast of the conventional explosive.

What is radiation?

Radiation is a form of energy that is present all around us. Some of the Earth's background radiation comes from naturally occurring radioactive elements from space, the soil, and the sun, as well as from man-made sources, like x-ray machines. Different types of radiation exist, some of which have more energy than others, and some of which can be more harmful than others. The dose of radiation that a person receives is measured in a unit called a *rem*. A rem is a measure of radiation dose, based on the amount of energy absorbed in a mass of tissue. For example, an average person gets about 1/3rd of a rem from exposure to natural sources of radiation in 1 year, and approximately 1/100th of a rem from one chest x-ray.

Are terrorists interested in radioactive materials?

Yes, terrorists have been interested in acquiring radioactive and nuclear material for use in attacks. For example, in 1995, Chechen extremists threatened to bundle radioactive material with explosives to use against Russia, in order to force the Russian military to withdraw from Chechnya. While no explosives were used, officials later retrieved a package of cesium-137 the rebels had buried in a Moscow park.

Since September 11, 2001, terrorist arrests and prosecutions overseas have revealed that individuals associated with al-Qaeda planned to acquire materials for an RDD. In 2004, British authorities arrested a British national, Dhiren Barot, and several associates on various charges, including conspiring to commit public nuisance by the use of radioactive materials. In 2006, Barot was found guilty and sentenced to life. British authorities disclosed that Barot developed a document known as the *Final Presentation*. The document outlined his research on the production of dirty bombs, which he characterized as designed to "cause injury, fear, terror and chaos" rather than to kill. U.S. federal prosecutors indicted Barot and two associates for conspiracy to use weapons of mass destruction against persons within the United States, in conjunction with the alleged surveillance of several suspects.

Will an RDD make me sick?

The effects of an RDD can vary, depending on what type of radioactive material is used and how much material is scattered. It is very difficult to design an RDD that would deliver large enough radiation doses to cause immediate health effects or fatalities in a large number of people. For the most part, an RDD would most likely be used to

- Contaminate facilities or places where people live and work, disrupting lives and livelihoods
- Cause anxiety in those who think they are being, or have been, exposed to radiation

How can I protect myself in a radiation emergency?

If an explosion occurs, it may not be known immediately that radioactive material is involved. If you are made aware that you are near the site of an RDD or potential release of radioactive material, you should

- Stay away from any obvious plume or dust cloud.
- Walk inside a building with closed doors and windows as quickly as possible and listen for alarms and further emergency response information.
- Cover your mouth and nose with a tissue, filter, clothing or damp cloth if there is dust in the air, to avoid inhaling or ingesting radioactive material.
- Remove contaminated clothing as soon as possible and place them in a sealed plastic bag. The clothing could be used later to estimate a person's exposure.
- Gently wash skin to remove any possible contamination, making sure that no radioactive material enters the mouth or is transferred to areas of the face where it could be easily moved to the mouth and swallowed.

If you are advised to take shelter, whether it is at home or in an office, you should

- Close all the doors and windows.
- Turn off ventilation, air conditioners, and forced air heating units that bring in fresh air from the outside. Only use units to re-circulate air that is already in the building.
- Close fireplace dampers.
- Move to an inner room.
- Keep your radio tuned to the emergency response network.

Is it necessary to purchase potassium iodide tablets for protection against radiation?

Potassium iodide (KI), which is available over the counter, protects people from thyroid cancer caused by radioactive iodine, a type of radioactive material that can be released in nuclear explosions, and depending on the amount released, can later cause thyroid cancer. KI should only be taken in a radiation emergency that involves the release of radioactive iodine. Since the use or release of radioactive iodine from an RDD is highly unlikely, KI pills would not be useful.

Conclusion

Terrorist attacks using a radiological dispersal device (RDD, also known as a "dirty bomb") carry the potential to shut down operations and cause substantial economic and psychological impacts. Risk and economic analysis methods may be employed to identify the most likely dirty bomb attack scenarios in terms of sources of radiological material, delivery modes, and detonation sites. The consequences of a successful attack are described in terms of human health effects and economic losses.

The chances of a successful dirty bomb attack are lower than expected, and the health consequences of even a major attack are relatively small. However, the economic consequences from a shutdown could result in significant losses. The implications of detecting, intercepting, and countering a dirty bomb attack must be considered a reality.

chapter eleven

Decontamination procedures

Definition of decontamination

Decontamination is the process of removing or neutralizing contaminants that have accumulated on personnel and equipment. This process is critical to health and safety at hazardous material response sites. Decontamination protects end users from hazardous substances that may contaminate and eventually permeate the protective clothing, respiratory equipment, tools, vehicles, and other equipment used in the vicinity of the chemical hazard; it protects all plant or site personnel by minimizing the transfer of harmful materials into clean areas; it helps prevent mixing of incompatible chemicals; and it protects the community by preventing uncontrolled transportation of contaminants from the site.

There are two types of decontamination:

- *Gross decontamination* to allow end users to safely exit or doff chemical protective clothing
- *Decontamination* and the reuse of chemical protective clothing

Prevention of contamination

The first step in decontamination is to establish standard operating procedures that minimize contact with chemicals and thus the potential for contamination. For example:

- Stress work practices that minimize contact with hazardous substances (e.g., do not walk through areas of obvious contamination; do not directly touch potentially hazardous substances).
- Use remote sampling, handling, and container-opening techniques (e.g., drum grapples, pneumatic impact wrenches).
- Protect monitoring and sampling instruments by bagging; make openings in the bags for sample ports and sensors that must contact site materials.
- Wear disposable outer garments and use disposable equipment where appropriate.

- Cover equipment and tools with a strippable coating that can be removed during decontamination.
- Encase the source of contaminants, for example with plastic sheeting or over packs.
- Ensure all closures and ensemble component interfaces are completely secured, and that no open pockets that could serve to collect contaminant are present.

Types of contamination

Surface Contaminants. Surface contaminants may be easy to detect and remove.

Permeated Contaminants. Contaminants that have permeated a material are difficult or impossible to detect and remove. If contaminants that have permeated a material are not removed by decontamination, they may continue to permeate the material, where they can cause an unexpected exposure.

Four major factors affect the extent of permeation:

1. *Contact time*. The longer a contaminant is in contact with an object, the greater the probability and extent of permeation. For this reason, minimizing contact time is one of the most important objectives of a decontamination program.
2. *Concentration*. Molecules flow from areas of high concentration to areas of low concentration. As concentrations of chemicals increase, the potential for permeation of personal protective clothing increases.
3. *Temperature*. An increase in temperature generally increases the permeation rate of contaminants.
4. *Physical state of chemicals*. As a rule, gases, vapors, and low-viscosity liquids tend to permeate more readily than high-viscosity liquids or solids.

Decontamination methods

Decontamination methods either (1) physically remove contaminants; (2) inactivate contaminants by chemical detoxification or disinfection/sterilization; or (3) remove contaminants by a combination of both physical and chemical means.

In general, gross decontamination is accomplished using detergents (surfactants) in water combined with a physical scrubbing action. This process will remove most forms of surface contamination including dusts, many inorganic chemicals, and some organic chemicals. Soapy water scrubbing of protective suits may not be effective in removing oily or tacky organic substances (e.g., PCBs in transformer oil). Furthermore, this form of decontamination is unlikely to remove any contamination that has permeated or penetrated the suit materials. Using organic solvents, such as petroleum distillates, may allow easier removal of heavy organic contamination but may result in other problems, including

- Permeation into clothing components, pulling the contaminant with it
- Spread of localized contaminant into other areas of the clothing
- Generation of large volumes of contaminated solvents that require disposal

One promising method for removing internal or matrix contamination is the forced circulation of heated air over clothing items for extended periods of time. This allows many organic chemicals to migrate out of the materials and evaporate into the heated air.

The process does require, however, that the contaminating chemicals be volatile. Additionally, low-level heat may accelerate the removal of plasticizer from garment materials and affect the adhesives involved in garment seams.

Unfortunately, both manufacturers and protective clothing authorities provide few specific recommendations for decontamination. There is no definitive list with specific methods recommended for specific chemicals and materials. Much depends on the individual chemical–material combination involved.

Testing the effectiveness of decontamination

Protective clothing or equipment reuse depends on demonstrating that adequate decontamination has taken place. Decontamination methods vary in their effectiveness, and unfortunately there are no completely accurate methods for nondestructively evaluating clothing or equipment contamination levels.

Methods which may assist in a determination include

- Visual examination of protective clothing for signs of discoloration, corrosive effects, or any degradation of external materials. However, many contaminants do not leave any visible evidence.
- Wipe sampling of external surfaces for subsequent analysis; this may or may not be effective for determining levels of surface contamination and depends heavily on the material–chemical combination. These methods will not detect permeated contamination.

- Evaluation of the cleaning solution. This method cannot quantify clean-method effectiveness, since the original contamination levels are unknown. The method can only show if chemical has been removed by the cleaning solution. If a number of garments have been contaminated, it may be advisable to sacrifice one garment for destructive testing by a qualified laboratory with analysis of contamination levels on and inside the garment.

Decontamination plan

1. A decontamination plan should be developed and set up before any personnel or equipment are allowed to enter areas where the potential for exposure to hazardous substances exists. The decontamination plan should
   a. Determine the number and layout of decontamination stations
   b. Determine the decontamination equipment needed
   c. Determine appropriate decontamination methods
   d. Establish procedures to prevent contamination of clean areas
   e. Establish methods and procedures to minimize wearer contact with contaminants during removal of personal protective clothing
   f. Establish methods for disposing of clothing and equipment that are not completely decontaminated
2. The plan should be revised whenever the type of personal protective clothing or equipment changes, the use conditions change, or the chemical hazards are reassessed based on new information.
3. The decontamination process should consist of a series of procedures performed in a specific sequence. For chemical protective ensembles, outer, more heavily contaminated items (e.g., outer boots and gloves) should be decontaminated and removed first, followed by decontamination and removal of inner, less contaminated items (e.g., jackets and pants). Each procedure should be performed at a separate station in order to prevent cross contamination. The sequence of stations is called the decontamination line.
4. Stations should be separated physically to prevent cross contamination and should be arranged in order of decreasing contamination, preferably in a straight line. Separate flow patterns and stations should be provided to isolate workers from different contamination zones containing incompatible wastes. Entry and exit points to exposed areas should be conspicuously marked. Dressing stations for entry to the decontamination area should be separate from redressing areas for exit from the decontamination area. Personnel who wish to enter clean areas of the decontamination facility, such as locker rooms, should be completely decontaminated.

5. All equipment used for decontamination must be decontaminated and/or disposed of properly. Buckets, brushes, clothing, tools, and other contaminated equipment should be collected, placed in containers, and labeled. Also, all spent solutions and wash water should be collected and disposed of properly. Clothing that is not completely decontaminated should be placed in plastic bags, pending further decontamination and/or disposal.
6. Decontamination of workers who initially come in contact with personnel and equipment leaving exposure or contamination areas will require more protection from contaminants than decontamination workers who are assigned to the last station in the decontamination line. In some cases, decontamination personnel should wear the same levels of protective clothing as workers in the exposure or contaminated areas. In other cases, decontamination personnel may be sufficiently protected by wearing one level lower protection (e.g., wearing Level B protection while decontaminating workers who are wearing Level A).

Decontamination for protective clothing reuse

Due to the difficulty in assessing contamination levels in chemical protective clothing before and after exposure, the responsible supervisor or safety professional must determine if the respective clothing can be reused. This decision involves considerable risk in determining clothing to be contaminant-free. Reuse can be considered if, in the estimate of the supervisor,

- No "significant" exposures have occurred
- Decontamination methods have been successful in reducing contamination levels to safe or acceptable concentrations

Contamination by known or suspected carcinogens should warrant automatic disposal. Use of disposable suits is highly recommended when extensive contamination is expected.

Emergency decontamination

1. In addition to routine decontamination procedures, emergency decontamination procedures must be established. In an emergency, the primary concern is to prevent the loss of life or severe injury to personnel. If immediate medical treatment is required to save a life, decontamination should be delayed until the victim is stabilized. If decontamination can be performed without interfering with essential life-saving techniques or first aid, or if a worker has

been contaminated with an extremely toxic or corrosive material that could cause severe injury or loss of life, decontamination should be continued.
2. If an emergency due to a heat-related illness develops, protective clothing should be removed from the victim as soon as possible to reduce the heat stress. During an emergency, provisions must also be made for protecting medical personnel and disposing of contaminated clothing and equipment.

Conclusion

There are many considerations and objectives involved in the management of large numbers of victims in chemical weapons of mass destruction (WMD) incidents. Many of these revolve around a primary objective, the transportation of the injured victims to a hospital. It is essential to decontaminate victims prior to transportation to stop further injury caused by contaminants, to eliminate the spread of lethal agents, and keep hospitals from closing because of contamination. The effectiveness of mass decontamination at a chemical agent attack will have a huge impact on the success of your operation.

Much of the mystery surrounding WMD operations comes from the misconception that a WMD incident is the "Mother of all Hazardous Materials Calls" and, as such, it is the prerogative of the hazardous material response teams (HMRTs) to handle most aspects of a successful mitigation. The simple truth is that HMRTs do not possess the resources necessary to deal with hundreds or thousands of victims. Nor will the HMRT arrive in time to take the actions necessary to save the lives of patients contaminated with a chemical agent. Even the largest metropolitan departments do not have the hazardous materials resources necessary to address thousands or even hundreds of victims of a chemical attack.

chapter twelve

Decontamination of chemical warfare and industrial agents

Introduction to decontamination methods and procedures

Decontamination is an important, unavoidable part of protection against chemical warfare agents. The aim of decontamination is to rapidly and effectively render harmless, or remove, poisonous substances, both on personnel and equipment. High decontamination capacity is one of the factors which may reduce the effect of an attack with chemical warfare (CW) agents. In this way, it may act as a deterrent.

The need for decontamination should be minimized to the extent possible by contamination avoidance and early warning. Equipment can be covered, for example, or easily decontaminated equipment can be chosen by means of suitable design and resistant surface cover.

Decontamination is time consuming and requires resources. Nerve agents and substances causing injury to the skin and tissue are easily soluble in, and penetrate many different types of, material, such as paint, plastics and rubber, rendering decontamination more difficult. If CW agents have penetrated sufficiently deep, then toxic gases can be released from the material for long periods. By adding substances which increase the viscosity of a CW agent, its persistence time and adhesive ability can be increased. These thickened agents will thus be more difficult to decontaminate with liquid decontaminants, since they adhere to the material and are difficult to dissolve.

The need for decontamination can only be established by means of detection. If detection is not possible, then decontamination must be done solely on suspicion of contamination.

Decontaminants

All decontamination is based on one or more of the following principles:

- To destroy CW agents by chemically modifying them (destruction)

- To physically remove CW agents by absorption, washing or evaporation
- To physically screen-off the CW agent so that it causes no damage

Most CW agents can be destroyed by means of suitable chemicals. Some chemicals are effective against practically all types of substances. However, such chemicals may be unsuitable for use in certain conditions since they corrode, etch, or erode the surface. Sodium hydroxide dissolved in organic solvent breaks down most substances but should not be used in decontaminating skin, other than in extreme emergencies when alternative means are not available.

Decontaminants that have effect only against a certain group of substances can be chosen as an alternative to a substance with general effect, on the condition that they will have a faster and better effect against the substance in question and/or a milder effect on the victim. Examples of such substances are chloramine solutions which are often used to decontaminate personnel. These have good effect against mustard agent and V-agents, but are ineffective against nerve agents of G-type (sarin, soman, tabun). A water solution of soda rapidly renders nerve agents of G-type harmless, but when used in connection with V-agents, it produces a final product which is almost as toxic as the original substance. This does not prevent V-agents being washed off with a soda solution, provided a sufficient amount is used. However, the final product will always be poisonous.

The disadvantage of specifically-acting decontaminants is partly that it is necessary to know which CW agent has been used and partly that access to several different types of decontaminating substances is required.

Decontamination methods

CW agents can be washed and rinsed away, dried up, sucked up by absorbent substances, or removed by heat treatment. Water, with or without additives of detergents, soda, soap, and so on, can be used, as well as organic solvents such as fuel, paraffin and carburetor spirit.

Emulsified solvents in water can be used to dissolve and wash off CW agents from various contaminated surfaces.

When decontaminating by washing, consideration must be taken of the poisonous substance remaining in the decontaminant, unless the CW agent has first been destroyed. The penetration ability of a CW agent can be enhanced when mixed with solvent. Today, there is an international development toward chemically resistant paints and materials, which implies that water-based methods will become more effective. However, the need for penetrating decontamination methods will remain for many years.

When washing with water—particularly with hot water and detergent—the CW agent will often be decomposed to some extent through hydrolysis. Detergents containing perforates are particularly effective in destroying nerve agents. Without an addition of perforates in the detergent, the hydrolysis products of V-agents may still remain toxic unless the pH is sufficiently high. Mustard agent is encapsulated by the detergent and, consequently, the hydrolysis rate decreases in comparison with clean water. However, the low solubility of mustard agent makes it difficult to remove without the addition of detergent, but the water used will still contain undestroyed mustard agent.

Small areas of terrain, for example, first-aid stations, may be decontaminated by removal of the top-soil. Another alternative is to cover the soil with chlorinated lime powder (sludge), which is a decontaminant with general effect and which releases active chlorine. CW agents which have penetrated into the soil, from where they release toxic vapor, are screened-off since the gas and liquid is destroyed by the chlorinated lime.

The physical screening-off of CW agents in the terrain by spreading a layer of soil or gravel over the contaminated area. The effect will be improved if bleaching powder is mixed into the covering material. Another example of covering is to use special plastic foil to cover contaminated areas inside vehicles. In this way, the personnel will be protected against transfer of liquid.

Individual decontamination

The most important decontamination measure naturally concerns the individual. If it is suspected that skin has been exposed to liquid CW agents, then it must be decontaminated immediately (within a minute). All experience confirms that the most important factor is time; the means used in decontamination are of minor importance. Good results can be obtained with such widely differing means as talcum powder, flour, soap and water, or special decontaminants.

In complete decontamination, clothes and personal belongings must also be decontaminated. If clothes have been exposed to liquid contamination, then extreme care must be taken when undressing to avoid transferring CW agents to the skin. There may be particular problems when caring for the injured, since it may be necessary to remove their clothes by cutting them off. This must be done in such a way that the patient is not further injured through skin contact with CW agents. During subsequent treatment it is essential to ensure that the entire patient is decontaminated, to avoid the risk of exposing the medical staff to the CW agents.

In most countries, such equipment includes means for individual decontamination, generally a mixture of chlorinated lime and magnesium oxide.

This decontamination process works by absorbing liquid substances and also by releasing free chlorine, which has a destructive effect on CW agents. The dry powder also has good effect on thickened agents since it bakes together the sticky substance, making it easier to remove. Personal decontaminants containing chlorinated lime have, however, an irritating effect on the skin. Consequently, comprehensive use should be followed by a bath or shower within a few hours.

Liquid personal decontaminants are common in some countries. Sodium phenolate or sodium cresolate in alcohol solution are used for individual decontamination of nerve agents. Chloramines in alcohol solution, possibly with additional substances, are commonly used against, for example, mustard agent. Instead of liquid individual decontaminants, it is possible to use an absorbent powder such as bentonite ("fuller's earth"). In the United States, the wet method formerly used was replaced by a decontaminant powder based on a mixture of resins, which decompose CW agents, and an absorbent.

A factor common to all individual decontaminants is that they can effectively remove CW agents on the surface of the skin. However, they have only limited ability to remove CW agents which have become absorbed by the skin, even very superficially. CW agents that have penetrated into the skin therefore function as a reservoir, which may further contribute to the poisoning, even after decontamination is completed.

In some cases, a wet method may give a better result in decontaminating deeply penetrated agents than a dry method. Reports from France indicate that a solution of potassium permanganate effectively destroys CW agents on the surface of the skin and also has a certain penetrating effect.

There are also individual decontaminants which can simultaneously function as a protective cream for use as a prophylactic. Canada has developed a mixture of a reactive substance (potassium 2, 3-butanedione monoximate) in polyethylene glycol, which has both these properties. It can be applied to the skin either as a cream or with a moist tissue.

Decontamination of equipment

Immediate decontamination of personal equipment and certain other kinds of smaller equipment is generally done with individual decontaminants. However, these substances are only capable of decontaminating liquid CW agents covering the surface. The decontamination is mainly done to prevent further penetration into the material and to decrease the risk when handling the equipment.

CW agents easily penetrate different materials and into crevasses, and will thus be difficult to reach by methods only designed for superficial decontamination. When a CW agent has penetrated into the surface, it is necessary to use some kind of deep-penetrating method. If such a method cannot be used, then it must be realized that the equipment cannot be used for a long period. Depending on the type of CW agent used and prevailing weather, that is, temperature, wind velocity, and precipitation (water solubility), the "self-decontamination" may take many days, or even weeks. The absorption into the surface and natural chemical degradation are important factors influencing the self-contamination period.

The diffusion and evaporation rate of CW agents from material is speeded-up considerably under warmer conditions. The decontamination tent used by the Swedish Army is heated with a mixture of hot exhaust gases and air from a small jet-pulse engine.

The tent is used for decontamination of lighter articles, for example, personal equipment. The decontamination container used by the civil defense forces is a development of the tent and heated with heat-exchanged hot air from a diesel burner.

The temperature in the tent is kept at about 130°C and in the container at 80°–130°C, depending on the type of material to be decontaminated. Decontamination time varies between two and five hours depending on the temperature.

Other methods utilize the heat of steam within a hot air stream which is blown against the contaminated surface. Decontamination by boiling is also an effective method. The advantage in comparison with heat is that hot water hydrolyzes and renders harmless many types of CW agents. The method may be of some interest in small-scale decontamination of rubber material, for example, protective masks.

Decontamination of CW agents which have penetrated deeply into the surface can also be done with decontaminants which are capable of penetrating the contaminated material. There are different substances with varying properties. A modern decontaminant is the German emulsion which consists of calcium hypochlorite, tetrachlorethylene, emulsifier ("phase transfer" catalyst), and water. Instead of tetrachlorethylene, the more environmentally harmless xylene is sometimes used.

Decontamination of vehicles and other large objects sometimes is done with steam and suspension and/or emulsion systems. A German company has developed special equipment, the C8-DADS (Direct Application Decontamination System), with which the emulsion is prepared and then dispersed onto the vehicle or the terrain.

Generally, it is an advantage to give the material an initial flushing with water before the chemical solution is added. A Swedish development of this approach is ongoing, where the intention is to spray water on, for example, a vehicle which passes through a flushing arch. The flushing arch has several jets which are supplied with water from a powerful pump. Another type of equipment which can use water from lakes, and so

Table 12.1 Decontamination contact time

Substance		No contact risk
	Liquid	Gas
Untreated metal surface		
Soman	<5 h	<5 h
Mustard agent	<20 h	<20 h
VX	6–8 days	6–8 days
Painted metal surface		
Soman	3–4 h	1–5 days
Mustard agent	1 day	3 days
VX	6 days	12–15 days

on, has been developed by a Norwegian firm. This is used for both flushing with cold and hot water/steam and also as a field shower.

In order to facilitate decontamination and decrease the risk when touched, the material can be painted with chemical-resistant paint systems, for example, polyurethane paint. Design of the equipment is also of major importance for ease of decontamination.

Example of self-decontamination times for contamination on metal surfaces and on a typical (non-resistant) paint at +15 °C, 4 m/s, and 2 mm large droplets

Note. The times for "liquid" only indicate when the surface is free of liquid, for example, no liquid is transferred when touched.

There is still a risk involved in contact and inhalation through release of gas from surfaces where the CW agent has penetrated deeply (Table 12.1).

Guidelines for mass casualty decontamination during a terrorist chemical agent incident

Relationships and process

The general process approach to dealing with a terrorist CW incident involves

- Providing technical solutions to specific issues on the efficacies and priorities of decontaminating people contaminated with chemical agents
- Identifying mass decontamination methods that can be readily applied, using existing resources, and that are consistent with current emergency responder procedures, training, logistic feasibility, and other potential considerations (human nature and psychology) and constraints (resource limits, civil liberty, environment preservation)

The general principles identified to guide emergency responder policies, procedures, and actions after a chemical agent incident:

- Decontaminate victims as soon as possible.
- Disrobing is decontamination; head to toe, more removal is better.
- Water flushing generally is the best mass decontamination method.
- After a known exposure to liquid chemical agent, emergency responders should be decontaminated as soon as possible to avoid serious effects.

Purposes of decontamination

The three most important reasons for decontaminating exposed victims are to

1. Reduce further possible agent exposure and further effects among victims by removing the agent from the victim's skin and clothing
2. Protect emergency responders and medical personnel from secondary transfer exposures
3. Provide victims with psychological comfort at, or near, the incident site, so as to prevent them from spreading contamination over greater areas

Rapid physical removal of agent from the victim is the single most important action associated with effective decontamination. Physical removal includes scraping or blotting off visible agent from the skin, disrobing, using adsorbents to soak up the agent, and flushing or showering with large quantities of water.

After a chemical agent attack, vapor or aerosol hazards may still be present, especially if the agent was disseminated within an enclosed structure. Furthermore, potentially toxic levels of chemical agent vapor may be trapped inside clothes and could continue to affect people, even after they leave the incident site.

Since the most important aspect of decontamination is the timely and effective removal of the agent, the precise methods used to remove the agent are not nearly as important as the speed at which the agent is removed. Scientific literature examining the effectiveness of different types of solutions in preventing chemical effects, and the widespread, ready availability of large quantities of water, determine that mass decontamination of large numbers of people can be most readily and effectively accomplished with a water shower system.

First responders may become contaminated during the conduct of decontamination operations. It is recommended that all responders participating in these procedures follow the *Medical Management of Chemical Casualties Handbook; October 2008*. United States Army guidance outlined in National Fire Protection Administration (NFPA) 471, *Recommended Practices for Response to Hazardous Materials Incidents*, states that

> "Even small amounts (several droplets) of liquid nerve agent contacting the unprotected skin can be severely incapacitating or lethal if the victim or responder is not decontaminated rapidly (within minutes) and treated medically."

Methods of mass decontamination

Decontamination must be conducted as soon as possible to save lives. Firefighters should use resources that are immediately available and start decontamination as soon as possible. Since they can bring large amounts of water to bear, the most expedient approach is to use currently available equipment to provide an emergency low-pressure deluge.

The following forms of water-based decontamination were considered:

- Water alone. Flushing or showering uses shear force and dilution to physically remove chemical agent from skin. Water alone is an excellent decontamination solution.
- Soap and water. By adding soap, a marginal improvement in results can be achieved by ionic degradation of the chemical agent. Soap aids in dissolving oily substances like mustard or blister agent. Liquid soaps are quicker to use than solids, and reduce the need for mechanical scrubbing; however, when scrubbing, potential victims should not abrade the skin. A disadvantage of soap is the need to have an adequate supply on hand. Additionally, extra time may be spent employing it, and using soap may hydrate the skin, possibly increasing damage by blister agents.
- Bleach and water. Bleach (sodium hypochlorite) and water solutions remove, hydrolyze, and neutralize most chemical agents. However, this approach is not recommended in a mass decontamination situation where speed is the paramount consideration for the following reasons:
  - Skin contact time is excessive. Laboratory studies show that chemical agents and relatively nontoxic, aqueous decontaminants may need to be in contact for durations longer than expected shower durations for significant reaction to occur.
  - Laboratory studies suggest that bleach solutions at the 0.5% level may not be better than flushing with water alone.
  - Medically, bleach solutions are not recommended for use near eyes or mucous membranes, or for those with abdominal, thoracic, or neural wounds.

In summary, the issues associated with the use of soap and bleach solutions include time delay, dilution and application, medical contraindications, and its efficacy compared to water. These limitations make the use of soap or bleach solutions less desirable than using water alone. The recommended rapid use of water, with or without soap, for the decontamination process should never be delayed to add soap or any other additive.

Decontamination procedures

Decontamination by removing clothes and flushing or showering with water is the most expedient and the most practical method for mass casualty decontamination. Disrobing and showering meets all the purposes and principles of decontamination. Showering is recommended whenever liquid transfer from clothing to skin is suspected. Disrobing should occur prior to showering for chemical agents; however, the decision to disrobe should be made by the Incident Commander based upon the situation. The wetting down of casualties as they start to disrobe speeds up the decontamination process, and is recommended for decontaminating biological or radiological casualties. However, this process may

- Force chemical agents through the clothing if water pressure is too high
- Decrease the potential efficacy of directly showering skin afforded by shear forces and dilution
- Relocate chemical agent within the actual showering area, thereby increasing the chance of contamination spread through personal contact and shower water runoff. The Mass Casualty Decontamination Research Team (MCDRT) recommends that victims remove clothing at least down to their undergarments prior to showering. Victims should be encouraged to remove as much clothing as possible, proceeding from head to toe. Victims unwilling to disrobe should shower clothed before leaving the decontamination area. It is also recommended that emergency responders use a high volume of water delivered at a minimum of 60 lb per square inch (psi) water pressure (standard household shower pressures usually average between 60 and 90 psi) to ensure the showering process physically removes viscous agent. Hypochlorite Solution as a Decontaminant in Sulfur Mustard Contaminated Skin.

The actual showering time will be an incident-specific decision, but may be as long as two to three minutes per individual under ideal situations. When large numbers of potential casualties are involved and queued for decontamination, showering time may be significantly shortened. This may also be dependent upon the volume of water available in the showering facilities.

In the course of decontaminating victims, first responders may inadvertently become contaminated. High pressure, low-volume decontamination showers are recommended primarily for wet decontamination of emergency responders in Level A suits after a HAZMAT incident. This gross decontamination procedure forcibly removes the contaminant from the personal protective equipment (PPE) worn by the emergency responders while conserving water. Often a secondary wash, and possible a tertiary wash, and rinse station are used.

Decontamination approaches (Figure 12.1)

Field-expedient water decontamination methods

Emergency responders should not overlook existing facilities when identifying means for rapid decontamination methods. For example,

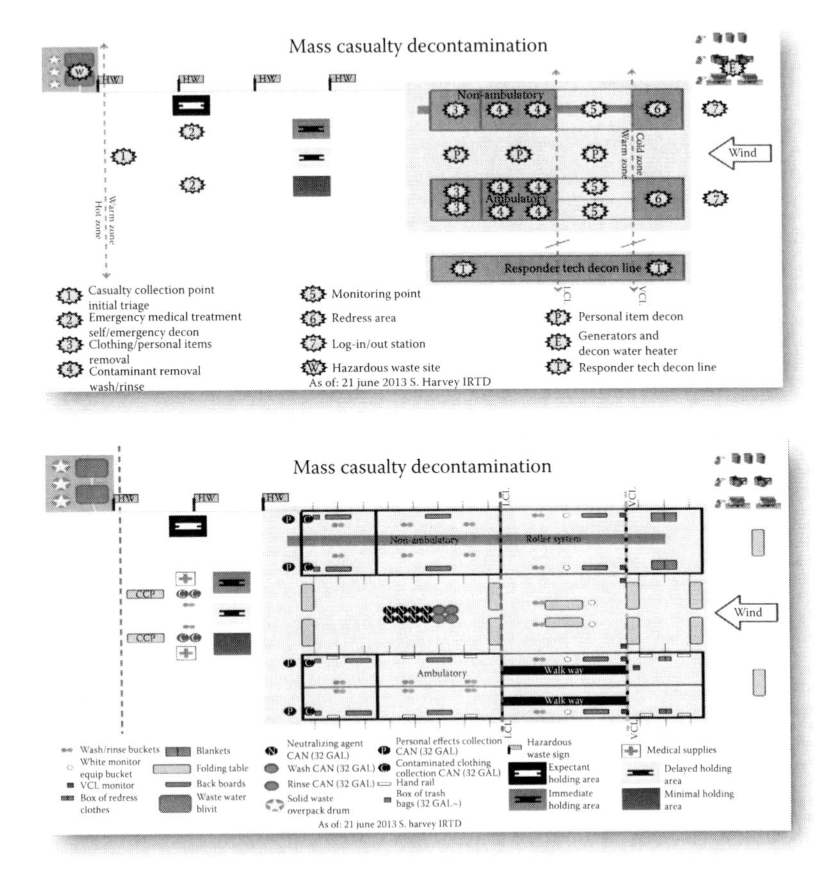

Figure 12.1 Schematic for mass decontamination.

although water damage to a facility might ensue, the necessity of saving victims' lives would justify the activation of overhead fire sprinklers for use as showers. Similarly, having victims wade and wash in water sources, such as public fountains, chlorinated swimming pools, swimming areas, and so on, provides an effective, high-volume decontamination technique.

Non-aqueous methods

The use of dry, gelled, or powdered decontaminating materials that adsorb the chemical agent are appropriate if their use is expedient. Commonly available absorbents include dirt, flour, fuller's earth, baking powder, sawdust, charcoal, ashes, activated carbon, alumina, silica gels, zeolites, clay materials, and tetra calcium aluminates. Although these absorbents may be expedient means of decontamination, their efficacy has not been determined.

Types of chemical victims

Three large-scale casualty events provide insight into the operational issues associated with casualty distribution and subsequent assessment that may be encountered during a response to a chemical terrorist incident. During Operation Desert Storm, 39 Iraqi Scud missiles reached the ground, with some landing in or around Tel Aviv, Israel.

The attacks resulted in approximately 1000 treated casualties with only two deaths. Even though it was never demonstrated that any of the Scuds contained chemical agents, the well-known possibility that the Scuds might contain chemical agents caused 544 anxiety attacks and 230 atropine overdoses. Approximately 75% of the overall casualties resulted from fears and reactions of the victims.

The second event occurred in Bhopal, India on 2–3 December 1984. During the night, several thousand gallons of highly volatile methylisocyanate was accidentally released over a three-hour period. This release was caused by the introduction of water into a methylisocyanate storage tank. The release resulted in over 200,000 people being exposed to the deadly gas, and as many as 5,000 died and over 60,000 were seriously and/or permanently injured.

The third event was the Japanese subway incident, where a reported 5510 victims sought medical treatment in 278 different hospitals and health clinics. Of the 5510 victims, 12 casualties died, 17 casualties were considered critically ill, 37 were casualties that were considered seriously ill, and 984 were casualties that were considered moderately ill.

Prioritizing casualties for decontamination

Decontamination prioritization describes the process of deciding the need for and order of victim decontamination. Triage is the medical process of prioritizing treatment urgency within a large group of victims. Both processes may be executed at the same time. The number of apparent victims from a chemical agent terrorist incident may exceed emergency responders' capabilities to effectively rescue, decontaminate, and treat victims, whether or not they have been exposed to chemical agent.

Responders, therefore, must prioritize victims for receiving decontamination, treatment, and medical evacuation, while providing the greatest benefit for the greatest number.

Although many emergency response services prepare for such incidents, few are currently capable of treating victims inside the hot zone. Therefore, whenever large numbers of victims are involved, it is recommended that they be sorted into ambulatory and non-ambulatory triage categories.

Prioritization for decontamination can effectively be performed in a manner that will maximize treatment while minimizing the number of emergency responders exposed to chemical agent.

Triage definitions

- Ambulatory casualties: Victims able to understand directions, talk, and walk unassisted. Most ambulatory victims are triaged as minimal (green tag/ribbon or Priority 3) unless severe signs/symptoms are present.
- Non-ambulatory casualties: Victims who are unconscious, unresponsive, or unable to move unassisted (Figure 12.2).

Ambulatory casualties should be directed to move upwind into an assembly area within the warm zone, where they can be prioritized for decontamination by onsite medical personnel. Factors that are recommended for determining the highest priority for ambulatory victim decontamination are highlighted in the picture representing an example of a mass casualty victim tag.

The highest priority for ambulatory decontamination are those casualties who were closest to the point of release and report they were exposed to an aerosol or mist, have some evidence of liquid deposition on clothing or skin or have serious medical symptoms (e.g., shortness of breath, chest tightness, etc.). The next priority are those ambulatory casualties who were not as close to the point of release, and may not have evidence of liquid deposition on clothing or skin, but who are clinically symptomatic. Victims suffering conventional injuries, especially open wounds, should be considered next. The lowest decontamination priority

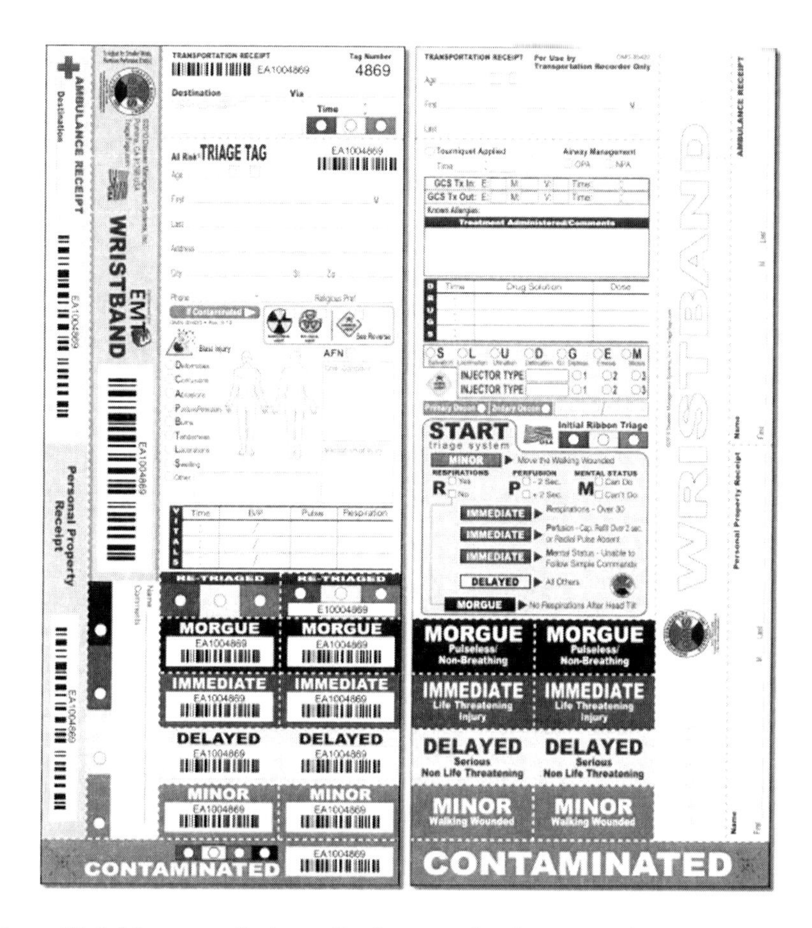

Figure 12.2 Mass casualty/casualty decontamination processing tags.

goes to ambulatory casualties who were far away from the point of release and who are asymptomatic.

Emergency responders should direct ambulatory victims in a prioritized fashion into the warm zone for decontamination. Care must be taken to ensure that the victims do not traverse contaminated areas in the hot zone or transfer contamination to the decontamination area.

Factors that determine highest priority for ambulatory victim decontamination

- Casualties closest to the point of release
- Casualties reporting exposure to vapor or aerosol
- Casualties with evidence of liquid deposition on clothing or skin
- Casualties with serious medical symptoms (shortness of breath, chest tightness, etc.)
- Casualties with conventional injuries

Non-ambulatory casualties may be more seriously injured than ambulatory victims and will remain in place while further prioritization for decontamination occurs.

It is recommended that prioritization of non-ambulatory victims for decontamination should be done using medical triage systems, such as START (Simple Triage and Rapid Treatment/Transport).

Mass casualty/casualty decontamination processing tag system

Priority: Immediate

Red tag 1: Respiration is present only after repositioning the airway. Applies to victims with respiratory rate > 30. Capillary refill delayed more than 2 s.

- Serious signs/symptoms
- Known liquid agent contamination

Priority: Delayed

Yellow tag 2: Victim displaying injuries that can be controlled/treated for a limited time in the field.

- Moderate to minimal signs/symptoms
- Known or suspected liquid agent
- Contamination
- Known aerosol contamination
- Close to point of release

Priority: Minor

Green tag 3: Ambulatory, with or without minor traumatic injuries that do not require immediate or significant treatment.

- Minimal signs/symptoms
- No known or suspected exposure to liquid, aerosol, or vapor

Priority: Deceased/expectant

Black tag 4: No spontaneous effective respiration present after an attempt to reposition the airway.

Casualty processing

The Incident Commander must quickly assess the scene and assign personnel to coordinate and manage both the medical triage and decontamination functions. If sufficient resources exist, two mass casualty decontamination systems should be established: one for ambulatory victims and one for non-ambulatory victims.

If available resources are only sufficient for a single system, non-ambulatory victims triaged as immediate are higher priority than the ambulatory victims triaged as immediate; therefore, they may be decontaminated as depicted in Figure 12.3. It is recommended that the remaining casualties should be processed in the same manner, with non-ambulatory victims being decontaminated before ambulatory victims. Due to the complex nature of some of these casualties (i.e., mixed chemical and conventional casualties), the medical triage and decontamination sectors should work closely together to maximize their collective sorting and management of casualties (Figure 12.3).

Additional considerations

Environmental concerns

The Environmental Protection Agency (EPA) has addressed the issues of acceptable levels of contamination in runoff and first responder liability for the spread of contamination caused by efforts to save lives. Regarding the liability issue, the EPA's interpretation of The Comprehensive Environmental Response, Compensation, and Liability Act (CERCLA) indicates that

> "No person shall be liable for costs or damages as a result of actions taken or omitted in the course of rendering care, assistance or advice in accordance with the National Contingency Plan (NCP) or at the direction of an On-Scene Coordinator appointed under such plan."

On the subject of accepted runoff, the EPA recognizes that any level of contamination represents a threat to the environment.

The threat is also dependent on many variables, including the involved chemicals, their concentrations, and the runoff watershed. However, life and health considerations are again paramount.

> "First responders should undertake any necessary emergency actions to save lives and protect the public and themselves. Once any imminent threats to human health and lives are addressed, first responders should immediately take all reasonable efforts to contain the contamination and avoid or mitigate environmental consequences."

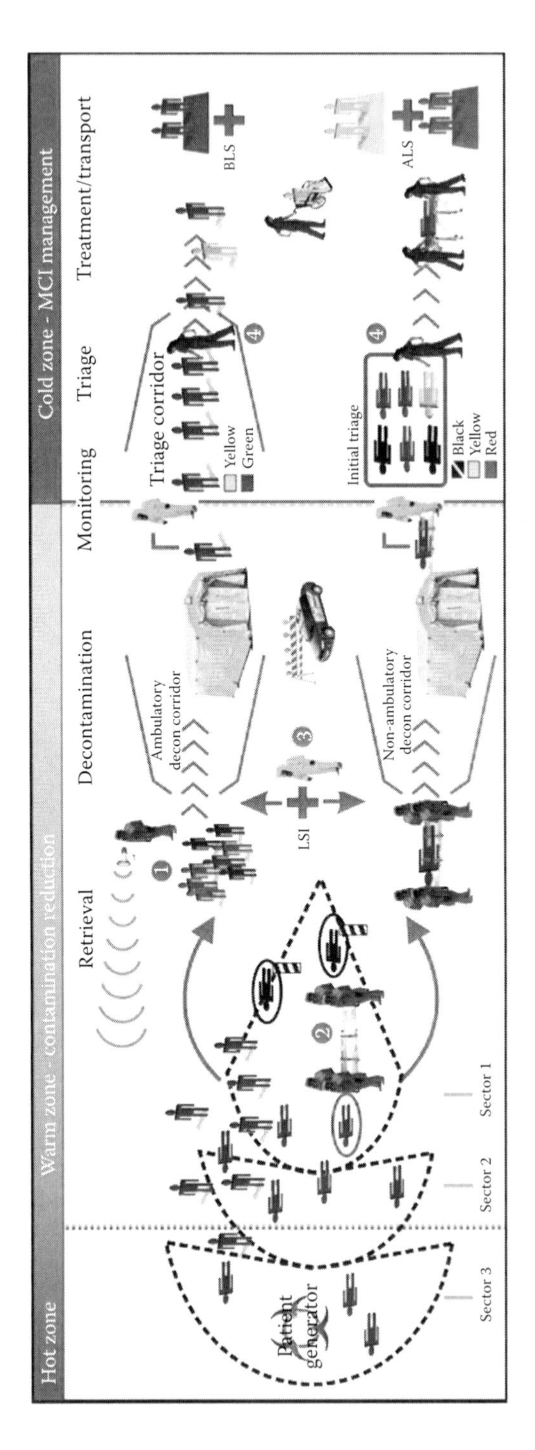

Figure 12.3 Decontamination flow.

Conclusion

There are many considerations and objectives involved in the management of large numbers of victims in a chemical weapons of mass destruction (WMD) incident. Many of these revolve around a primary objective, the transportation of the injured victims to a hospital. It is essential to decontaminate victims prior to transportation to stop further injury caused by contaminants, to eliminate the spread of lethal agents, and keep hospitals from closing because of contamination. The effectiveness of mass decontamination at a chemical agent attack will have a huge impact on the success of your operation.

Much of the mystery surrounding WMD operations comes from the misconception that a WMD incident is the "Mother of all Hazardous Materials Calls" and, as such, it is the prerogative of the hazardous material response teams (HMRT) to handle most aspects of a successful mitigation. The simple truth is that HMRTs do not possess the resources necessary to deal with hundreds or thousands of victims. Nor will the HMRT arrive in time to take the actions necessary to save the lives of patients contaminated with a chemical agent.

Even the largest metropolitan departments do not have the hazardous materials resources necessary to address thousands or even hundreds of victims of a chemical attack.

chapter thirteen

Chemical protective clothing

The purpose of chemical protective clothing and equipment is to shield or isolate individuals from the chemical, physical, and biological hazards that may be encountered during hazardous materials operations. During chemical operations, it is not always apparent when exposure occurs. Many chemicals pose invisible hazards and offer no warning properties.

These guidelines describe the various types of clothing that are appropriate for use in various chemical operations, and provide recommendations in their selection and use.

The final paragraph discusses heat stress and other key physiological factors that must be considered in connection with protective clothing use.

It is important that protective clothing users realize that no single combination of protective equipment and clothing is capable of protecting them against all hazards. Thus, protective clothing should be used in conjunction with other protective methods. The use of protective clothing can itself create significant wearer hazards, such as heat, physical, and psychological stress, in addition to impaired vision, mobility, and communication. The greater the level of chemical protective clothing the greater the associated risks for any given situation. Equipment and clothing should be selected that provide an adequate level of protection. Overprotection as well as under-protection can be hazardous and should be avoided.

Protective clothing applications

Protective clothing must be worn whenever the wearer faces potential hazards arising from chemical exposure. Some examples include

- Emergency response
- Chemical manufacturing and process industries
- Hazardous waste site cleanup and disposal
- Asbestos removal and other particulate operations
- Agricultural application of pesticides

Within each application, there are several operations which require chemical protective clothing. For example, in emergency response, the following activities require chemical protective clothing use.

Site survey: The initial investigation of a hazardous materials incident; these situations are usually characterized by a large degree of uncertainty and mandate the highest levels of protection.

Rescue: Entering a hazardous materials area for the purpose of removing an exposure victim; special considerations must be given to how the selected protective clothing may affect the ability of the wearer to carry out rescue and to the contamination of the victim.

Spill mitigation: Entering a hazardous materials area to prevent a potential spill or to reduce the hazards from an existing spill (e.g., applying a chlorine kit on railroad tank car). Protective clothing must accommodate the required tasks without sacrificing adequate protection.

Emergency monitoring: Outfitting personnel in protective clothing for the primary purpose of observing a hazardous materials incident without entry into the spill site. This may be applied to monitoring contract activity for spill cleanup.

Decontamination: Applying decontamination procedures to personnel or equipment leaving the site; in general, a lower level of protective clothing is used by personnel involved in decontamination.

Clothing ensemble: The approach in selecting personal protective clothing must encompass an "ensemble" of clothing and equipment items which are easily integrated to provide both an appropriate level of protection and still allow one to carry out activities involving chemicals. In many cases, simple protective clothing by itself may be sufficient to prevent chemical exposure, such as wearing gloves in combination with a splash apron and face shield (or safety goggles).

The following is a checklist of components that may form the chemical protective ensemble:

- Protective clothing (suit, coveralls, hoods, gloves, boots)
- Respiratory equipment (SCBA [self-contained breathing apparatus], combination SCBA/SAR [supplied air respirator], air purifying respirators)
- Cooling system (ice vest, air circulation, water circulation)
- Communications device
- Head protection
- Eye protection
- Ear protection
- Inner garment
- Outer protection (over gloves, over boots, flash cover)

Factors that affect the selection of ensemble components include

- How each item accommodates the integration of other ensemble components. Some ensemble components may be incompatible due

to how they are worn (e.g., some SCBAs may not fit within a particular chemical protective suit or allow acceptable mobility when worn).
- The ease of interfacing ensemble components without sacrificing required performance (e.g., a poorly fitting over glove that greatly reduces wearer dexterity).
- Limiting the number of equipment items to reduce donning time and complexity (e.g., some communications devices are built into SCBAs which as units are NIOSH certified).

Level of protection

Table 13.1 lists ensemble components based on the widely used *EPA Levels of Protection: Levels A, B, C, and D*. These lists can be used as the starting point for ensemble creation; however, each ensemble must be tailored to the specific situation in order to provide the most appropriate level of protection. For example, if an emergency response activity involves a highly contaminated area or if the potential for contamination is high, it may be advisable to wear a disposable covering, such as Tyvek coveralls or PVC splash suits, over the protective ensemble.

The type of equipment used and the overall level of protection should be re-evaluated periodically as the amount of information about the chemical situation or process increases, and when workers are required to perform different tasks. Personnel should upgrade or downgrade their level of protection only in concurrence with the site supervisor, safety officer, or plant industrial hygienist.

The recommendations in Table 13.1 serve only as guidelines. It is important to realize that selecting items by how they are designed or configured alone is not sufficient to ensure adequate protection. In other words, just having the right components to form an ensemble is not enough. The EPA levels of protection do not define what performance the selected clothing or equipment must offer. Many of these considerations are described in the "limiting criteria" column of Table 13.1. Additional factors relevant to the various clothing and equipment items are described in subsequent paragraphs.

Ensemble selection factors

- *Chemical hazards*. Chemicals present a variety of hazards such as toxicity, corrosiveness, flammability, reactivity, and oxygen deficiency. Depending on the chemicals present, any combination of hazards may exist.
- *Physical environment*. Chemical exposure can happen anywhere: in industrial settings, on the highways, or in residential areas. It may occur either indoors or outdoors; the environment may be extremely hot, cold, or moderate; the exposure site may be relatively uncluttered or rugged, presenting a number of physical hazards; chemical handling activities may involve entering confined spaces, heavy

Table 13.1 EPA levels of protection

Level A

Vapor-protective suit (meets NFPA 1991)
Pressure-demand, full-face SCBA
Inner chemical-resistant gloves, chemical-resistant safety boots, two-way radio communication
Optional: Cooling system, outer gloves, hard hat
Protection provided: Highest available level of respiratory, skin, and eye protection from solid, liquid, and gaseous chemicals.
Used when: The chemical(s) have been identified and present high-level hazards to respiratory system, skin and eyes. Substances are present which are known or suspected to be toxic to skin or carcinogenic. Operations must be conducted in confined or poorly ventilated areas.
Limitations: Protective clothing must resist permeation by the chemical or mixtures present. Ensemble items must allow integration without loss of performance.

Level B

Liquid splash-protective suit (meets NFPA 1992)
Pressure-demand, full-face piece SCBA
Inner chemical-resistant gloves, chemical-resistant safety boots, two-way radio communications
Hard hat.
Optional: Cooling system, outer gloves
Protection provided: Provides same level of respiratory protection as Level A, but less skin protection. Liquid splash protection, but no protection against chemical vapors or gases.
Used when: The chemical(s) have been identified but do not require a high level of skin protection. Initial site surveys are required until higher levels of hazards are identified. The primary hazards associated with site entry are from liquid and not vapor contact.
Limitations: Protective clothing items must resist penetration by the chemicals or mixtures present. Ensemble items must allow integration without loss of performance.

Level C

Support function protective garment (meets NFPA 1993)
Full-face piece, air-purifying, canister-equipped respirator
Chemical resistant gloves and safety boots
Two-way communications system, hard hat
Optional: Face shield, escape SCBA
Protection provided: The same level of skin protection as Level B, but a lower level of respiratory protection. Liquid splash protection but no protection to chemical vapors or gases.

(*Continued*)

Table 13.1 (Continued) EPA levels of protection

Used when: Contact with site chemical(s) will not affect the skin. Air contaminants have been identified and concentrations measured. A canister is available which can remove the contaminant. The site and its hazards have been completely characterized.
Limitations: Protective clothing items must resist penetration by the chemical or mixtures present. Chemical airborne concentration must be less than IDLH levels. The atmosphere must contain at least 19.5% oxygen.
Not acceptable for chemical emergency response

Level D

Coveralls, safety boots/shoes, safety glasses or chemical splash goggles
Optional: Gloves, escape SCBA, face-shield
Protection provided: No respiratory protection, minimal skin protection.
Used when: The atmosphere contains no known hazard. Work functions preclude splashes, immersion, potential for inhalation, or direct contact with hazard chemicals.
Limitations: This level should not be worn in the hot zone. The atmosphere must contain at least 19.5% oxygen.
Not acceptable for chemical emergency response

lifting, climbing a ladder, or crawling on the ground. The choice of ensemble components must account for these conditions.

- *Duration of exposure.* The protective qualities of ensemble components may be limited to certain exposure levels (e.g., material chemical resistance, air supply). The decision for ensemble use time must be made assuming the worst-case exposure so that safety margins can be applied to increase the protection available to the worker.
- *Protective clothing or equipment available.* Hopefully, an array of different clothing or equipment is available to workers to meet all intended applications. Reliance on one particular clothing or equipment item may severely limit a facility's ability to handle a broad range of chemical exposures. In its acquisition of equipment and clothing, the safety department or other responsible authority should attempt to provide a high degree of flexibility while choosing protective clothing and equipment that is easily integrated and provides protection against each conceivable hazard.

Classification of protective clothing

Personal protective clothing includes the following:

- Fully encapsulating suits
- Non-encapsulating suits

- Gloves, boots, and hoods
- Firefighter's protective clothing
- Proximity, or approach, clothing
- Blast or fragmentation suits
- Radiation-protective suits

Firefighter turnout clothing, proximity gear, blast suits, and radiation suits by themselves are not acceptable for providing adequate protection from hazardous chemicals.

Various types of protection clothing available are described in the type of protection they offer, and lists factors to consider in their selection and use (Table 13.2).

Classification of chemical protective clothing. A listing of clothing classifications is provided in 1–3. Clothing can be classified by design, performance, and service life (Table 13.3).

Design. Categorizing clothing by design is mainly a means for describing what areas of the body the clothing item is intended to protect.

In emergency response, hazardous waste site cleanup, and dangerous chemical operations, the only acceptable types of protective clothing include fully or totally encapsulating suits and non-encapsulating, or "splash," suits, plus accessory clothing items such as chemically resistant gloves or boots. These descriptions apply to how the clothing is designed and not to its performance.

Performance: The National Fire Protection Association (NFPA) has classified suits by their performance:

a. Vapor-protective suits (NFPA Standard 1991) provide "gas-tight" integrity and are intended for response situations where no chemical contact is permissible. This type of suit would be equivalent to the clothing required in the EPA's Level A.
b. Liquid splash-protective suits (NFPA Standard 1992) offer protection against liquid chemicals in the form of splashes, but not against continuous liquid contact or chemical vapors or gases. Essentially, this type of clothing would meet the EPA Level B needs. It is important to note, however, that by wearing liquid splash-protective clothing, the wearer accepts exposure to chemical vapors or gases because this clothing does not offer gas-tight performance. The use of duct tape to seal clothing interfaces does not provide the type of wearer encapsulation necessary for protection against vapors or gases.
c. Support function protective garments (NFPA Standard 1993) must also provide liquid splash protection but offer limited physical protection. These garments may comprise several separate protective clothing components (i.e., coveralls, hoods, gloves, and boots). They are intended for use in nonemergency, nonflammable situations

Table 13.2 Types of protective clothing for full body protection

Description	Type of protection	Use considerations
Fully encapsulating suit One-piece garment. Boots and gloves may be integral, attached and replaceable, or separate.	Protects against splashes, dust gases, and vapors.	Does not allow body heat to escape. May contribute to heat stress in wearer, particularly if worn in conjunction with a closed-circuit SCBA; a cooling garment may be needed. Impairs worker mobility, vision, and communication.
Non-encapsulating suit Jacket, hood, pants or bib overalls, and one-piece coveralls.	Protects against splashes, dust, and other materials but not against gases and vapors. Does not protect parts of head or neck.	Do not use where gas-tight or pervasive splashing protection is required. May contribute to heat stress in wearer. Tape-seal connections between pant cuffs and boots and between gloves and sleeves.
Aprons, leggings, and sleeve protectors Fully sleeved and gloved apron. Separate coverings for arms and legs. Commonly worn over non-encapsulating suit.	Provides additional splash protection of chest, forearms, and legs.	Whenever possible, should be used over a non-encapsulating suit to minimize potential heat stress. Useful for sampling, labeling, and analysis operations. Should be used only when there is a low probability of total body contact with contaminants.
Firefighters' protective clothing Gloves, helmet, running or bunker coat, running or bunker pants (NFPA No. 1971, 1972, 1973, and boots [1974]).	Protects against heat, hot water, and some particles. Does not protect against gases and vapors, or chemical permeation or degradation. NFPA Standard No. 1971 specifies that a garment consists of an outer shell,	Decontamination is difficult. Should not be worn in areas where protection against gases, vapors, chemical splashes or permeation is required.

(Continued)

Table 13.2 (Continued) Types of protective clothing for full body protection

Description	Type of protection	Use considerations
	An inner liner and a vapor barrier with a minimum water penetration of 25 lb. /in.^{2} (1.8 kg/cm^2) to prevent passage of hot water.	
Proximity garment (approach suit) One- or two-piece over garment with boot covers, gloves, and hood of aluminized nylon or cotton fabric. Normally worn over other protective clothing, firefighters' bunker gear, or flame-retardant coveralls.	Protects against splashes, dust, gases, and vapors.	Does not allow body heat to escape. May contribute to heat stress in wearer, particularly if worn in conjunction with a closed-circuit SCBA; a cooling garment may be needed. Impairs worker mobility, vision, and communication.
Blast and fragmentation suit Blast and fragmentation vests and clothing, bomb blankets, and bomb carriers.	Provides some protection against very small detonations. Bomb blankets and baskets can help redirect a blast.	Does not provide for hearing protection.
Radiation-contamination protective suit Various types of protective clothing designed to prevent contamination of the body by radioactive particles.	Protects against alpha and beta particles. Does *not* protect against gamma radiation.	Designed to prevent skin contamination. If radiation is detected on site, consult an experienced radiation expert and evacuate personnel until the radiation hazard has been evaluated.
Flame/fire retardant coveralls Normally worn as an undergarment.	Provides protection from flash fires.	Adds bulk and may exacerbate heat stress problems and impair mobility.

where the chemical hazards have been completely characterized. Examples of support functions include proximity to chemical processes, decontamination, hazardous waste clean up, and training. Support function protective garments should not be used in chemical emergency response or in situations where chemical hazards remain uncharacterized.

Table 13.3 Classification of chemical protective clothing

By design	By performance	By service life
Gloves	Particulate protection	Single use
Boots	Liquid splash protection	Limited use
Aprons, jackets, coveralls, full body suits	Vapor protection	Reusable

These NFPA standards define minimum performance requirements for the manufacture of chemical protective suits. Each standard requires rigorous testing of the suit and the materials that comprise the suit in terms of overall protection, chemical resistance, and physical properties.

Suits that are found compliant by an independent certification and testing organization may be labeled by the manufacturer as meeting the requirements of the respective NFPA standard. Manufacturers also have to supply documentation showing all test results and characteristics of their protective suits.

Protective clothing should completely cover both the wearer and his or her breathing apparatus. In general, respiratory protective equipment is not designed to resist chemical contamination. Level A protection (vapor-protective suits) requires this configuration. Level B ensembles may be configured either with the SCBA on the outside or inside. However, it is strongly recommended that the wearer's respiratory equipment be worn inside the ensemble to prevent its failure and to reduce decontamination problems. Level C ensembles use cartridge or canister type respirators which are generally worn outside the clothing.

Service life of protective clothing

Clothing item service life is an end-user decision depending on the costs and risks associated with clothing decontamination and reuse. For example, a Saranex/Tyvek garment may be designed to be a coverall (covering the wearer's torso, arms, and legs) intended for liquid splash protection, which is disposable after a single use.

Protective clothing may be labeled as (a) reusable, for multiple wearing; or (b) disposable, for one-time use.

The distinctions between these types of clothing are both vague and complicated. Disposable clothing is generally lightweight and inexpensive. Reusable clothing is often more rugged and costly. Nevertheless, extensive contamination of any garment may render it disposable. The basis of this classification really depends on the costs involved in purchasing, maintaining, and reusing protective clothing versus the alternative of disposal following exposure. If an end user can anticipate obtaining several uses out of a garment while still maintaining adequate protection from that garment at lower cost than its disposal, the suit becomes reusable. Yet, the key assumption in this determination is the viability of the garment following exposure. This issue is further discussed in the paragraph on decontamination.

Protective clothing selection factors

Clothing design: Manufacturers sell clothing in a variety of styles and configurations. Design considerations include

- Clothing configuration
- Components and options
- Sizes
- Ease of donning and doffing
- Clothing construction
- Accommodation of other selected ensemble equipment
- Comfort
- Restriction of mobility

Material chemical resistance: Ideally, the chosen material(s) must resist permeation, degradation, and penetration by the respective chemicals.

Permeation is the process by which a chemical dissolves in or moves through a material on a molecular basis. In most cases, there will be no visible evidence of chemicals permeating a material.

Permeation breakthrough time is the most common value used to assess material chemical compatibility. The rate of permeation is a function of several factors, such as chemical concentration, material thickness, humidity, temperature, and pressure. Most material testing is done with 100% chemical over an extended exposure period.

The time it takes chemical to permeate through the material is the breakthrough time. An acceptable material is one where the breakthrough time exceeds the expected period of garment use. However, temperature and pressure effects may enhance permeation and reduce the magnitude of this safety factor. For example, small increases in ambient temperature can significantly reduce breakthrough time and the protective barrier properties of a protective clothing material.

Degradation involves physical changes in a material as the result of a chemical exposure, use, or ambient conditions (e.g., sunlight). The most common signs of material degradation are discoloration, swelling, loss of physical strength, or deterioration.

Penetration is the movement of chemicals through zippers, seams, or imperfections in a protective clothing material.

It is important to note that no material protects against all chemicals or combinations of chemicals, and that no currently available material is an effective barrier to any prolonged chemical exposure.

Sources of information

Guidelines for the Selection of Chemical Protective Clothing, 3rd Edition. This reference provides a matrix of clothing material recommendations for approximately 500 chemicals based on an evaluation of chemical resistance test data, vendor literature, and raw material suppliers. The major limitation for these guidelines is their presentation of recommendations by generic material class. Numerous test results have shown that similar materials from different manufacturers may give widely different performance.

That is to say, manufacturer A's butyl rubber glove may protect against chemical X, but a butyl glove made by manufacturer B may not.

Quick Selection Guide to Chemical Protective Clothing. Pocket-size guide that provides chemical resistance data and recommendations for 11 generic materials against over 400 chemicals. The guide is color-coded by material-chemical recommendation. As with the "Guidelines" above, the major limitation of this reference is its dependence on generic data.

Vendor data or recommendations: The best source of current information on material compatibility should be available from the manufacturer of the selected clothing. Many vendors supply charts which show actual test data or their own recommendations for specific chemicals. However, *unless vendor data or the recommendations are well documented, end users must approach this information with caution.* Material recommendations must be based on data obtained from tests performed to standard ASTM methods. Simple ratings of "poor," "good," or "excellent" give no indication of how the material may perform against various chemicals.

Mixtures of chemicals can be significantly more aggressive toward protective clothing materials than any single chemical alone. One permeating chemical may pull another with it through the material. Very little data is available for chemical mixtures. Other situations may involve unidentified substances. In both cases of mixtures and unknowns, serious consideration must be given to deciding which protective clothing is selected. If clothing must be used without test data, garments with materials having the broadest chemical resistance should be worn, that is, materials which demonstrate the best chemical resistance against the widest range of chemicals.

Physical properties

As with chemical resistance, manufacturer materials offer wide ranges of physical qualities in terms of strength, resistance to physical hazards, and operation in extreme environmental conditions. Comprehensive manufacturing standards, such as the NFPA standards, set specific limits on these material properties, but only for limited applications, that is, emergency response.

End users in other applications may assess material physical properties by posing the following questions:

- Does the material have sufficient strength to withstand the physical strength of the tasks at hand?
- Will the material resist tears, punctures, cuts, and abrasions?
- Will the material withstand repeated use after contamination and decontamination?
- Is the material flexible or pliable enough to allow end users to perform needed tasks?
- Will the material maintain its protective integrity and flexibility under hot and cold extremes?
- Is the material flame-resistant or self-extinguishing (if these hazards are present)?
- Are garment seams in the clothing constructed so they provide the same physical integrity as the garment material?

Ease of decontamination: The degree of difficulty in decontaminating protective clothing may dictate whether disposable or reusable clothing is used, or a combination of both.

Cost: Protective clothing end users must endeavor to obtain the broadest protective equipment they can buy with available resources to meet their specific application.

Chemical protective clothing standards: Protective clothing buyers may wish to specify clothing that meets specific standards, such as 1910.120 or the NFPA standards (see paragraph on classification by performance). The NFPA standards do not apply to all forms of protective clothing and applications.

General guidelines

Decide if the clothing item is intended to provide vapor, liquid splash, or particulate protection

Vapor-protective suits also provide liquid splash and particulate protection. Liquid splash-protective garments also provide particulate protection. Many garments may be labeled as totally encapsulating but do not provide gas-tight integrity due to inadequate seams or closures. Gas-tight integrity can only be determined by performing a pressure or inflation test and a leak detection test of the respective protective suit. This test involves

- Closing off suit exhalation valves
- Inflating the suit to a pre-specified pressure
- Observing whether the suit holds the above pressure for a designated period

ASTM Standard Practice F1052 (1987 Edition) offers a procedure for conducting this test.

Splash suits must still cover the entire body when combined with the respirator, gloves, and boots. Applying duct tape to a splash suit does not make it protect against vapors. Particulate protective suits may not need to cover the entire body, depending on the hazards posed by the particulate. In general, gloves, boots, and some form of face protection are required.

Clothing items may only be needed to cover a limited area of the body such as gloves on hands. The nature of the hazards and the expected exposure will determine if clothing should provide partial or full body protection.

Determine if the clothing item provides full body protection

- Vapor-protective or totally encapsulating suit will meet this requirement by passing gas-tight integrity tests.
- Liquid splash-protective suits are generally sold incomplete (i.e., fewer gloves and boots).
- Missing clothing items must be obtained separately and match or exceed the performance of the garment.
- Buying a PVC glove for a PVC splash suit does not mean that you obtain the same level of protection. This determination must be made by comparing chemical resistance data.

Evaluate manufacturer chemical resistance data provided with the clothing
Manufacturers of vapor-protective suits should provide permeation resistance data for their products, while liquid and particulate penetration resistance data should accompany liquid splash and particulate protective garments, respectively. Ideally, data should be provided for every primary material in the suit or clothing item. For suits, this includes the garment, visor, gloves, boots, and seams.

Permeation factors—data should include the following:

- Chemical name
- Breakthrough time (shows how soon the chemical permeates)
- Permeation rate (shows the rate that the chemical comes through)
- System sensitivity (allows comparison of test results from different laboratories)
- A citation that the data was obtained in accordance with ASTM Standard Test Method F739-85

If no data are provided or if the data lack any one of the above items, the manufacturer should be asked to supply the missing data. Manufacturers that provide only numerical or qualitative ratings must support their recommendations with complete test data.

Liquid penetration data should include a pass or fail determination for each chemical listed, and a citation that testing was conducted in accordance with ASTM Standard Test Method F903-86. Protective suits which are certified to NFPA 1991 or NFPA 1992 will meet all of the above requirements.

Particulate penetration data should show some measure of material efficiency in preventing particulate penetration in terms of particulate type or size and percentage held out. Unfortunately, no standard tests are available in this area and end users may have little basis for comparing products.

Suit materials which show no breakthrough or no penetration to a large number of chemicals are likely to have a broad range of chemical resistance. (Breakthrough times greater than 1 hour are usually considered to be an indication of acceptable performance.) Manufacturers should provide data on the ASTM Standard Guide F1001-86 chemicals.

The fifteen liquid and six gaseous chemicals listed in Table 13.4 represent a cross-section of different chemical classes and challenges for protective clothing materials. Manufacturers should also provide test data on other chemicals. If there are specific chemicals within your operating area that have not been tested, ask the manufacturer for test data on these chemicals.

Clothing donning, doffing, and use

The procedures below are given for vapor-protective or liquid splash-protective suit ensembles and should be included in the training program.

Donning the ensemble

- A routine should be established and practiced periodically for donning the various ensemble configurations that a facility or team may use. Assistance should be provided for donning and doffing since these operations are difficult to perform alone, and solo efforts may increase the possibility of ensemble damage.
- Table 13.5 lists sample procedures for donning a totally encapsulating suit/SCBA ensemble. These procedures should be modified depending on the suit and accessory equipment used. The procedures assume the wearer has previous training in respirator use and decontamination procedures.
- Once the equipment has been donned, its fit should be evaluated. If the clothing is too small, it will restrict movement, increase the likelihood of tearing the suit material, and accelerate wearer fatigue.

Table 13.4 Sample donning procedures

1. Inspect clothing and respiratory equipment before donning (see paragraph on Inspection).
2. Adjust hard hat or headpiece, if worn, to fit user's head.
3. Open back closure used to change air tank (if suit has one) before donning suit.
4. Standing or sitting, step into the legs of the suit; ensure proper placement of the feet within the suit; then gather the suit around the waist.
5. Put on chemical-resistant safety boots over the feet of the suit. Tape the leg cuff over the tops of the boots.
   If additional chemical-resistant safety boots are required, put these on now. Some one-piece suits have heavy-soled protective feet. With these suits, wear short, chemical-resistant safety boots inside the suit.
6. Put on air tank and harness assembly of the SCBA. Don the face piece and adjust it to be secure, but comfortable. Do not connect the breathing hose. Open valve on air tank.
7. Perform negative and positive respirator face piece seal test procedures.
   To conduct a negative-pressure test, close the inlet part with the palm of the hand or squeeze the breathing tube so it does not pass air, and gently inhale for about 10 seconds. Any inward rushing of air indicates a poor fit. Note that a leaking face piece may be drawn tightly to the face to form a good seal, giving a false indication of adequate fit.
   To conduct a positive-pressure test, gently exhale while covering the exhalation valve to ensure that a positive pressure can be built up. Failure to build a positive pressure indicates a poor fit.
8. Depending on type of suit:
   i. Put on long-sleeved inner gloves (similar to surgical gloves). Secure gloves to sleeves, for suits with detachable gloves (if not done prior to entering the suit).
   ii. Additional over gloves, worn over attached suit gloves, may be donned later.
9. Put sleeves of suit over arms as assistant pulls suit up and over the SCBA. Have assistant adjust suit around SCBA and shoulders to ensure unrestricted motion.
10. Put on hard hat, if needed.
11. Raise hood overhead carefully so as not to disrupt face seal of SCBA mask. Adjust hood to give satisfactory comfort.
12. Begin to secure the suit by closing all fasteners on opening until there is only adequate room to connect the breathing hose. Secure all belts and/or adjustable leg, head, and waistbands.
13. Connect the breathing hose while opening the main valve.
14. Have assistant first ensure that wearer is breathing properly and then make final closure of the suit.
15. Have assistant check all closures.
16. Have assistant observe the wearer for a period of time to ensure that the wearer is comfortable, psychologically stable, and that the equipment is functioning properly.

Table 13.5 Sample doffing procedures

If sufficient air supply is available to allow appropriate decontamination before removal

17. Remove any extraneous or disposable clothing, boot covers, outer gloves, and tape.
18. Have assistant loosen and remove the wearer's safety shoes or boots.
19. Have assistant open the suit completely and lift the hood over the head of the wearer and rest it on top of the SCBA tank.
20. Remove arms, one at a time, from suit. Once arms are free, have assistant lift the suit up and away from the SCBA backpack—avoiding any contact between the outside surface of the suit and the wearer's body—and lay the suit out flat behind the wearer. Leave internal gloves on, if any.
21. Sitting, if possible, remove both legs from the suit.
22. Follow procedure for doffing SCBA.
23. After suit is removed, remove internal gloves by rolling them off the hand, inside out.
24. Remove internal clothing and thoroughly cleanse the body.

If the low-pressure warning alarm has sounded, signifying that approximately 5 minutes of air remain

25. Remove disposable clothing.
26. Quickly scrub and hose off, especially around the entrance/exit zipper.
27. Open the zipper enough to allow access to the regulator and breathing hose.
28. Immediately attach an appropriate canister to the breathing hose (the type and fittings should be predetermined). Although this provides some protection against any contamination still present, it voids the certification of the unit.
29. Follow Steps 1 through 8 of the regular doffing procedure above. Take extra care to avoid contaminating the assistant and the wearer.

If the clothing is too large, the possibility of snagging the material is increased, and the dexterity and coordination of the wearer may be compromised. In either case, the wearer should be recalled and better-fitting clothing provided.

Doffing an ensemble

- Exact procedures for removing a totally encapsulating suit/SCBA ensemble must be established and followed in order to prevent contaminant migration from the response scene and transfer of contaminants to the wearer's body, the doffing assistant, and others.
- Sample doffing procedures are provided in Table 13.5. These procedures should be performed only after decontamination of the suited

end user. They require a suitably attired assistant. Throughout the procedures, both wearer and assistant should avoid any direct contact with the outside surface of the suit.

User monitoring and training

The wearer must understand all aspects of clothing/equipment operation and their limitations; this is especially important for fully encapsulating ensembles where misuse could potentially result in suffocation. During protective clothing use, end users should be encouraged to report any perceived problems or difficulties to their supervisor. These malfunctions include, but are not limited to

- Degradation of the protection ensemble
- Perception of odors
- Skin irritation
- Unusual residues on clothing material
- Discomfort
- Resistance to breathing
- Fatigue due to respirator use
- Interference with vision or communication
- Restriction of movement
- Physiological responses such as rapid pulse, nausea, or chest pain

Before end users undertake any activity in their chemical protective ensembles, the anticipated duration of use should be established. Several factors limit the length of a mission, including

- Air supply consumption as affected by wearer work rate, fitness, body size, and breathing patterns
- Suit ensemble permeation, degradation, and penetration by chemical contaminants, including expected leakage through suit or respirator exhaust valves (ensemble protection factor)
- Ambient temperature as it influences material chemical resistance and flexibility, suit and respirator exhaust valve performance, and wearer heat stress
- Coolant supply (if necessary).

Inspection, storage, and maintenance

The end user, in donning protective clothing and equipment, must take all necessary steps to ensure that the protective ensemble will perform as expected. During emergencies is not the right time to discover discrepancies in the protective clothing. Teach end-user care for clothing and other protective equipment in the same manner as parachutists care for parachutes. Following a standard program for inspection, proper storage, and maintenance, along with realizing protective clothing/equipment limitations, is the best way to avoid chemical exposure during emergency response.

Inspection

An effective chemical protective clothing inspection program should feature five different inspections:

- Inspection and operational testing of equipment received as new from the factory or distributor
- Inspection of equipment as it is selected for a particular chemical operation
- Inspection of equipment after use or training and prior to maintenance
- Periodic inspection of stored equipment
- Periodic inspection when a question arises concerning the appropriateness of selected equipment, or when problems with similar equipment are discovered

Each inspection will cover different areas with varying degrees of detail. Those personnel responsible for clothing inspection should follow manufacturer directions; many vendors provide detailed inspection procedures. The generic inspection checklist provided may serve as an initial guide for developing more extensive procedures.

Records must be kept of all inspection procedures. Individual identification numbers should be assigned to all reusable pieces of equipment (many clothing and equipment items may already have serial numbers), and records should be maintained by that number. At a minimum, each inspection should record

- Clothing/equipment item ID number
- Date of the inspection
- Person making the inspection
- Results of the inspection
- Any unusual conditions noted

Periodic review of these records can provide an indication of protective clothing which requires excessive maintenance and can also serve to identify clothing that is susceptible to failure.

Storage

Clothing must be stored properly to prevent damage or malfunction from exposure to dust, moisture, sunlight, damaging chemicals, extreme temperatures, and impact. Procedures are needed for both initial receipt of equipment and after use or exposure of that equipment. Many manufacturers specify recommended procedures for storing their products. These should be followed to avoid equipment failure resulting from improper storage.

Some guidelines for general storage of chemical protective clothing include

- Potentially contaminated clothing should be stored in an area separate from street clothing or unused protective clothing.
- Potentially contaminated clothing should be stored in a well-ventilated area, with good air flow around each item, if possible.
- Different types and materials of clothing and gloves should be stored separately to prevent issuing the wrong material by mistake (e.g., many glove materials are black and cannot be identified by appearance alone).
- Protective clothing should be folded or hung in accordance with manufacturer instructions.

Maintenance

Manufacturers frequently restrict the sale of certain protective suit parts to individuals or groups who are specially trained, equipped, or authorized by the manufacturer to purchase them. Explicit procedures should be adopted to ensure that the appropriate level of maintenance is performed only by those individuals who have this specialized training and equipment. In no case should you attempt to repair equipment without checking with the person in your facility that is responsible for chemical protective clothing maintenance.

The following classification scheme is recommended to divide the types of permissible or non-permissible repairs.

- *Level 1*: User or wearer maintenance, requiring a few common tools or no tools at all.
- *Level 2*: Maintenance that can be performed by the response team's maintenance shop, if adequately equipped and trained.
- *Level 3*: Specialized maintenance that can be performed only by the factory or an authorized repair person.

Each facility should adopt the above scheme and list which repairs fall into each category for each type of protective clothing and equipment. Many manufacturers will also indicate which repairs, if performed in the field, void the warranty of their products. All repairs made must be recorded on the records for the specific clothing along with appropriate inspection results.

Training

Benefits of training in the use of protective clothing:

- Allows the user to become familiar with the equipment in a nonhazardous, nonemergency condition
- Instills confidence in the user in his/her equipment
- Makes the user aware of the limitations and capabilities of the equipment
- Increases worker efficiency in performing various tasks
- Reduces the likelihood of accidents during chemical operations

Content: Training should be completed prior to actual clothing use, in a non-hazardous environment, and should be repeated at the frequency required by OSHA SARA III legislation. As a minimum, the training should point out the user's responsibilities and explain the following, using both classroom and field training when necessary:

- The proper use and maintenance of selected protective clothing, including capabilities and limitations
- The nature of the hazards and the consequences of not using the protective clothing
- The human factors influencing protective clothing performance
- Instructions in inspecting, donning, checking, fitting, and using protective clothing
- Use of protective clothing in normal air for a long familiarity period
- The user's responsibility (if any) for decontamination, cleaning, maintenance, and repair of protective clothing
- Emergency procedures and self-rescue in the event of protective clothing/ equipment failure
- The buddy system

The discomfort and inconvenience of wearing chemical protective clothing and equipment can create a resistance to its conscientious use. One essential aspect of training is to make the user aware of the need for protective clothing and to instill motivation for the proper use and maintenance of that protective clothing.

Risks

Heat stress

Wearing full body chemical protective clothing puts the wearer at considerable risk of developing heat stress. This can result in health effects ranging from transient heat fatigue to serious illness or death. Heat stress is caused by a number of interacting factors, including:

- Environmental conditions
- Type of protective ensemble worn
- The work activity required
- The individual characteristics of the responder

When selecting chemical protective clothing and equipment, each item's benefit should be carefully evaluated for its potential for increasing the risk of heat stress. For example, if a lighter, less insulating suit can be worn without a sacrifice in protection, then it should be. Because the incidence of heat stress depends on a variety of factors, all workers wearing full body chemical protective ensembles should be monitored.

Review Paragraph III: Chapter 4, "Heat stress," in the *OSHA Technical Manual*. The following physiological factors should be monitored.

Heart rate

Count the radial pulse during a 30-second period as early as possible in any rest period. If the heart rate exceeds 110 beats per minute at the beginning of the rest period, the next work cycle should be shortened by one-third.

Oral temperature

- *Do not permit an end user to wear protective clothing and engage in work when his or her oral temperature exceeds 100.6°F (38.1°C).*
- Use a clinical thermometer (three minutes under the tongue) or similar device to measure oral temperature at the end of the work period (before drinking).
- If the oral temperature exceeds 99.6°F (37.6°C), shorten the next work period by at least one-third.
- If the oral temperature exceeds 99.6°F (37.6°C) at the beginning of a response period, shorten the mission time by one-third.

Body water loss

Measure the end user's weight on a scale accurate to plus or minus 0.25 pounds prior to any response activity. Compare this weight with his or her normal body weight to determine if enough fluids have been consumed to prevent dehydration. Weights should be taken while the end user wears similar clothing, or ideally, in the nude. The body water loss should not exceed 1.5% of the total body weight loss from a response.

Conclusion

Chemical protective clothing has been designed and developed to suit the requirements of chemical protection along with providing comfort.

The purpose of chemical protective clothing and equipment is to shield or isolate individuals from the chemical, physical, and biological hazards that may be encountered during hazardous materials operations. During chemical operations, it is not always apparent when exposure occurs. Many chemicals pose invisible hazards and offer no warning.

These guidelines describe the various types of clothing that are appropriate for use in various chemical operations, and provide recommendations in their selection and use. The final paragraph discusses heat stress and other key physiological factors that must be considered in connection with protective clothing use.

It is important that protective clothing users realize that no single combination of protective equipment and clothing is capable of protecting against all hazards. Thus, protective clothing should be used in conjunction with other protective methods. For example, engineering or administrative controls to limit chemical contact with personnel should always be considered as an alternative measure for preventing chemical exposure. The use of protective clothing can itself create significant wearer hazards, such as heat, physical, and psychological stress, in addition to impaired vision, mobility, and communication. In general, the greater the level of chemical protective clothing, the greater the associated risks. For any given situation, equipment and clothing should be selected that provides an adequate level of protection. Overprotection as well as under-protection can be hazardous and should be avoided.

chapter fourteen

Emergency response guidebook

The Emergency Response Guidebook: A Guidebook for First Responders during the Initial Phase of a Dangerous Goods/Hazardous Materials Transportation Incident (ERG) is used by emergency response personnel (such as firefighters and police officers) in Canada, Mexico, and the United States when responding to a transportation emergency involving hazardous materials. First responders in Argentina, Brazil, and Colombia have recently begun using the ERG as well. It is produced by the United States Department of Transportation, Transport Canada, and the Secretariat of Communications and Transportation (Mexico) (Figure 14.1).

Guidebook contents

The ERG is primarily applicable for hazardous materials transported by highway and railway, but is also applicable for materials transported by air or waterway, as well as by pipeline. It was first issued by the U.S. Department of Transportation in 1973, but later became a joint publication of the Department of Transportation (US DOT), Transport Canada (TC), and the Secretariat of Communications and Transportation (SCT) of Mexico, in collaboration with the Chemistry Information Center for Emergencies (CIQUIME) of Argentina.

The ERG is issued every four years, with editions now being published in Spanish (*Guía de Respuesta en Caso de Emergencia*). In 1996, it was published as the North American Emergency Response Guidebook, but by the next publication in 2000, "North American" was removed due to its use by several South American countries).

The ERG "is primarily a guide to aid first responders in quickly identifying the specific or generic hazards of the material(s) involved in the incident, and protecting themselves and the general public," and should only be used for the "initial response phase of the incident" ("that period following arrival at the scene of an incident during which the presence and/or identification of dangerous goods is confirmed, protective actions and area securement are initiated, and assistance of qualified personnel is requested"). It is divided into six color-coded sections (white [uncolored], yellow, blue, orange, green, and a second white [uncolored]).

The ERG includes 62 "guides" (found in the orange section) that identify the primary hazards associated with the applicable general category

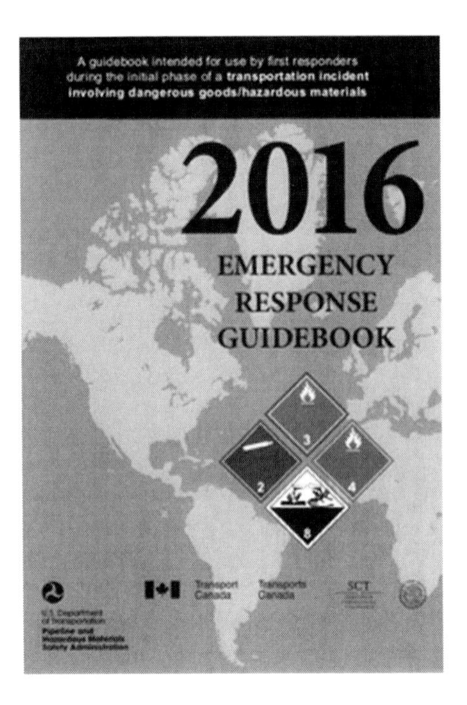

Figure 14.1 ERG.

of hazardous material and general guidance on how to respond to incidents involving that general category of hazardous material. The primary purpose of the ERG is to direct the emergency responders to the most appropriate of these guides, based on the incident. The ERG also provides guidance regarding recommended evacuation distances, if applicable, in the green section.

White section (front)

The first section, with white page (uncolored) borders, provides the following:

- Information regarding shipping documents
- Instructions on how to use the guidebook
- General guidance for responding to any hazardous material incident
- Basic information on the hazard classification system and the associated placards/labels
- Recommendations on the proper guides, based on transporting vehicle types and/or placards (when the material in question cannot be further identified otherwise)

- General safety precautions
- Specific guidance for incidents involving pipelines

Yellow section

The second section, with yellow page borders, references materials in order of their assigned four-digit ID number/UN/NA number (which is often placarded with the other hazardous materials placards) and identifies the appropriate guide number to reference in the orange section). Items highlighted in green in this section will have evacuation distances included in the green section.

Blue section

The third section, with blue page borders, references materials in alphabetical order and identifies the appropriate guide number to reference in the orange section). Items highlighted in green in this section will also have evacuation distances included in the green section.

Orange section

The fourth section, with orange page borders, includes the actual response guides. Each of the 62 guides provides safety recommendations and directions on how to proceed during the initial response phase (first thirty minutes) of the incident. It includes "health" and "fire or explosion" potential hazard information (with the most dangerous hazard listed first). For example, "the material gives off irritating vapors, easily ignited by heat, reactive with water," "highly toxic, may be fatal if inhaled, swallowed or absorbed through skin," and so on.

Next this section includes information for responders on appropriate protective clothing and possible evacuation information for either spill or fire. It also includes information on fighting fires (for example, do not apply water to sodium), warnings for spills or leaks, and special directions for first aid (for example, not to give mouth-to-mouth resuscitation if the materials are toxic).

In the event of an unknown material, Guide #111 should be followed until more information becomes available.

Green section

The fifth section, with green page borders, suggests initial evacuation or shelter-in-place distances (protective action distances) for spills of materials that are toxic by inhalation (TIH). These distances vary based on the size of the spill (small or large) and whether the incident occurs during the day or at night. Only materials that were highlighted in green in the yellow and blue sections are included in the green section. This section also includes information regarding toxic gases that are produced when certain materials are spilled in water (as identified previously in this section). Finally, this section includes some very specific evacuation details for six common materials.

White section (back)

The sixth section, with white page (uncolored) borders, provides the following:

- Additional instructions on how to use the guidebook
- Information regarding protective clothing and equipment
- Instructions on fire and spill control
- BLEVE (boiling liquid expanding vapor explosion) safety precautions
- Beginning with the 2004 edition, information specifically for hazardous materials being used for terrorism
- Glossary of terms used in the ERG
- Contact information for the various countries

Shipping list

A shipping list, packing list, waybill, packing slip (also known as a bill of parcel, unpacking note, packaging slip, (delivery) docket, delivery list, manifest or customer receipt), is a shipping document that accompanies delivery packages, usually inside an attached shipping pouch or inside the package itself. It commonly includes an itemized detail of the package contents and does not include customer pricing. It serves to inform all parties, including transport agencies, government authorities, and customers, about the contents of the package. It helps them deal with the package accordingly.

Contents

- Recipients
- Shipping list details
- See also
- References

Recipients

An invoice goes to the person responsible for paying the bill while the shipping list (or packaging slip) goes to the recipient. The shipping list is included in the shipped box. In some scenarios, the same person will pay the bill and receive the shipment. However, a person can buy and pay for a product and send it to someone else.

Shipping list details

Shipping lists vary depending on the business and its products. Every shipment to a customer should contain a shipping list that includes the order date, the products included within the box, and the quantity of each product. Some businesses may want to include the weight of the product next to the item. Many receivers use the shipping list as a guide or checklist when unpacking their order. If something is missing from the box, they cross-check it with the shipping list and then alert the seller.

Classification and labeling summary tables

Information on this graphic changes depending on which "division" of explosive is shipped. Explosive dangerous goods have compatibility group letters assigned to facilitate segregation during transport. The letters used range from A to S, excluding the letters I, M, O, P, Q, and R. The example shown features an explosive with compatibility group "A" (shown as 1.1A).

The actual letter shown would depend on the specific properties of the substance being transported.

Class 1: Explosives

The Canadian Transportation of Dangerous Goods Regulations provide a description of compatibility groups

- Explosives with a mass explosion hazard (TNT, dynamite, nitroglycerine)
- Explosives with a severe projection hazard
- Explosives with a fire, blast, or projection hazard but not a mass explosion hazard
- Minor fire or projection hazard (includes ammunition and most consumer fireworks)
- An insensitive substance with a mass explosion hazard (explosion similar to 1.1)
- Extremely insensitive articles

The United States Department of Transportation (DOT) regulates HAZMAT transportation within the territory of the United States

- Explosives with a mass explosion hazard (nitroglycerin/dynamite)
- Explosives with a blast/projection hazard

- Explosives with a minor blast hazard (rocket propellant, display fireworks)
- Explosives with a major fire hazard (consumer fireworks, ammunition)
- Blasting agents
- Extremely insensitive explosives

Class 2: Gases

Gases which are compressed, liquefied, or dissolved under pressure as detailed below. Some gases have subsidiary risk classes; poisonous or corrosive.

- 2.1 Flammable gas: Gases which ignite on contact with an ignition source, such as acetylene, hydrogen, and propane.
- 2.2 Nonflammable gases: Gases which are neither flammable nor poisonous. Includes the cryogenic gases/liquids (temperatures of below −100°C) used for cryopreservation, and rocket fuels, such as nitrogen, neon, and carbon dioxide.
- 2.3 Poisonous gases: Gases liable to cause death or serious injury to human health if inhaled; examples are fluorine, chlorine, and hydrogen cyanide.

Class 3: Flammable liquids

Class 3 flammable liquids are included in one of the following packing groups:

- Packing Group I, if they have an initial boiling point of 35°C or less at an absolute pressure of 101.3 kPa and any flash point, e.g., diethyl ether or carbon disulfide
- Packing Group II, if they have an initial boiling point greater than 35°C at an absolute pressure of 101.3 kPa and a flash point less than 23°C, such as gasoline (petrol) and acetone
- Packing Group III, if the criteria for inclusion in Packing Group I or II are not met, e.g., kerosene and diesel.

Class 4: Flammable solids

- 4.1 Flammable solids: Solid substances that are easily ignited and readily combustible (nitrocellulose, magnesium, safety or strike-anywhere matches)
- 4.2 Spontaneously combustible solids: Solid substances that ignite spontaneously (aluminum alkyls, white phosphorus).
- 4.3 Dangerous when wet: Solid substances that emit a flammable gas when wet, or react violently with water (sodium, calcium, potassium, calcium carbide).

Class 5: Oxidizing agents and organic peroxides

- 5.1 Oxidizing agents other than organic peroxides (calcium hypochlorite, ammonium nitrate, hydrogen peroxide, potassium permanganate)
- 5.2 Organic peroxide oxidizing agent, either in liquid or solid form (benzoyl peroxides, cumene hydroperoxide)

Class 6: Toxic and infectious substances

- 6.1a Toxic substances which are liable to cause death or serious injury to human health if inhaled, swallowed, or by skin absorption (potassium cyanide, mercuric chloride)
- 6.1b (Now PGIII) Toxic substances which are harmful to human health (N.B. this symbol is no longer authorized by the United Nations) (pesticides, methylene chloride)
- 6.2 Biohazardous substances; the World Health Organization (WHO) divides this class into two categories: Category A—infectious, and Category B—samples (virus cultures, pathology specimens, used intravenous needles).

Class 7: Radioactive substances

Radioactive substances comprise substances or a combination of substances which emit ionizing radiation (uranium, plutonium)

Class 8: Corrosive substances

Corrosive substances are substances that can dissolve organic tissue or severely corrode certain metals:

- 8.1 Acids: sulfuric acid, hydrochloric acid
- 8.2 Alkalis: potassium hydroxide, sodium hydroxide

Class 9: Miscellaneous

- Hazardous substances that do not fall into the other categories (asbestos, air-bag inflators, self-inflating life rafts, dry ice)

Boiling liquid expanding vapor explosion (BLEVE)

A boiling liquid expanding vapor explosion (BLEVE, /ˈblɛviː/ BLEV-ee) is an explosion caused by the rupture of a vessel containing a pressurized liquid above its boiling point.

Mechanism

There are three characteristics of liquids which are relevant to the discussion of a BLEVE. If a liquid in a sealed container is boiled, the pressure inside the container increases. As the liquid changes to a gas it expands—this expansion in a vented container would cause the gas and liquid to take up more space. In a sealed container, the gas and liquid are not able to take up more space and so the pressure rises. Pressurized vessels containing liquids can reach an equilibrium where the liquid stops boiling and the pressure stops rising. This occurs when no more heat is being added to the system (either because it has reached ambient temperature or has had a heat source removed).

The boiling temperature of a liquid is dependent on pressure—high pressures will yield high boiling temperatures, and low pressures will yield low boiling temperatures. A common, simple experiment is to place a cup of water in a vacuum chamber, and then reduce the pressure in the chamber until the water boils.

By reducing the pressure the water will boil even at room temperature. This works both ways—if the pressure is increased beyond normal atmospheric pressures, the boiling of hot water could be suppressed far beyond normal temperatures. The cooling system of a modern internal combustion engine is a real-world example.

When a liquid boils it turns into a gas. The resulting gas takes up far more space than the liquid did.

Typically, a BLEVE starts with a container of liquid which is held above its normal, atmospheric-pressure boiling temperature. Many substances normally stored as liquids, such as CO_2, propane, and other similar industrial gases have boiling temperatures, at atmospheric pressure, far below room temperature. In the case of water, a BLEVE could occur if a pressurized chamber of water was heated far beyond the standard 100°C (212°F). That container, because the boiling water pressurizes it, is capable of holding liquid water at very high temperatures.

If the pressurized vessel, containing liquid at high temperature (which may be room temperature, depending on the substance) ruptures, the pressure which prevents the liquid from boiling is lost. If the rupture is catastrophic, where the vessel is immediately incapable of holding any pressure at all, then there suddenly exists a large mass of liquid which is at very high temperature and very low pressure.

This causes a portion of the liquid to "instantaneously" boil, which in turn causes an extremely rapid expansion. Depending on temperatures, pressures, and the substance involved, that expansion may be so rapid that it can be classified as an explosion, fully capable of inflicting severe damage on its surroundings.

Water example

For example, a tank of pressurized liquid water held at 204.4°C (400°F) might be pressurized to 1.7 MPa (250 psi) above atmospheric ("gauge") pressure. If the tank containing the water were to rupture, there would for a brief moment exist a volume of liquid water which would be at

- Atmospheric pressure, and a
- Temperature of 204.4°C (400°F).

At atmospheric pressure the boiling point of water is 100°C (212°F), meaning liquid water at atmospheric pressure does not exist at temperatures higher than this. At that moment, the water would boil and turn to vapor explosively, and the 204.4°C (400°F) water, turned to gas, would take up significantly more volume (~22-fold) than it did as liquid, causing a vapor explosion. Such explosions can happen when the superheated water of a steam engine escapes through a crack in a boiler, causing a boiler explosion.

BLEVEs without chemical reactions

A BLEVE need not be a chemical explosion—nor does there need to be a fire—however, if a flammable substance is subject to a BLEVE it may also be subject to intense heating, either from an external source of heat which may have caused the vessel to rupture in the first place or from an internal source of localized heating, such as skin friction.

This heating can cause a flammable substance to ignite, adding a secondary explosion caused by the primary BLEVE. While blast effects of any BLEVE can be devastating, a flammable substance such as propane can add significantly to the danger.

While the term BLEVE is most often used to describe the results of a container of flammable liquid rupturing due to fire, a BLEVE can occur even with a non-flammable substance, such as water, liquid nitrogen, liquid helium, or other refrigerants or cryogens, and therefore is not usually considered a type of chemical explosion.

Fires

BLEVEs can be caused by an external fire near the storage vessel causing heating of the contents and pressure build-up. While tanks are often designed to withstand great pressure, constant heating can cause the metal to weaken and eventually fail. If the tank is being heated in an area where there is no liquid to absorb the heat, it may rupture faster. Gas containers are usually equipped with relief valves that vent off excess pressure, but the tank can still fail if the pressure is not released quickly enough.

Relief valves are sized to release pressure fast enough to prevent the pressure from increasing beyond the strength of the vessel, but not so fast as to be the cause of an explosion. An appropriately sized relief valve will allow the liquid inside to boil slowly, maintaining a constant pressure in the vessel until all the liquid has boiled and the vessel empties.

If the substance involved is flammable, it is likely that the resulting cloud of the substance will ignite after the BLEVE has occurred, forming a fireball and possibly a fuel-air explosion, also termed a *vapor cloud explosion* (VCE). If the materials are toxic, a large area will be contaminated.

Globally Harmonized System of Classification and Labeling of Chemicals

The Globally Harmonized System of Classification and Labeling of Chemicals (GHS) is an internationally agreed-upon system, created by the United Nations in 1992 and, as of 2015, not yet fully implemented in many countries. It was designed to replace the various classification and labeling standards used in different countries by using consistent criteria on a global level. It supersedes the relevant system of the European Union, which has incorporated the United Nations' GHS into EU law as the CLP Regulation. The United States Occupational Safety and Health Administration standards also implement the GHS.

Hazard classification

The GHS classification system is complex, with data obtained from tests, literature, and practical experience. The main elements of the hazard classification criteria are summarized below.

Physical hazards

Substances or articles are assigned to nine different hazard classes, largely based on the United Nations Dangerous Goods System. Additions and changes have been necessary, since the scope of the GHS includes all target audiences.

Explosives are assigned to one of six subcategories, depending on the type of hazard they present, as used in the UN Dangerous Goods System.

Gases are Category 1 flammable if they start to flame in a range in air at 20°C and a standard pressure of 101.3 kPa. Category 2 is nonflammable and nontoxic gases, and Category 3 is toxic gases. Substances and mixtures of this hazard class are assigned to one of two hazard categories on the basis of the outcome of the test or calculation method.

A flammable liquid is a liquid with a flash point of not more than 93°C. Substances and mixtures of this hazard class are assigned to one of four hazard categories on the basis of the flash point and boiling point. A pyrophoric liquid is a liquid that, even in small quantities, is liable to ignite within 5 min after coming into contact with air. Substances and mixtures of this hazard class are assigned to a single hazard category on the basis of the outcome of the UN Test N.3.

A flammable solid is one that is readily combustible or may cause or contribute to fire through friction. Readily combustible solids are powdered, granular, or pasty substances which are dangerous if they can be easily ignited by brief contact with an ignition source, such as a burning match, and if the flame spreads rapidly. They are further divided into

- Flammable solids
- Polymerizing substances
- Self-reactive substances; that is, thermally unstable solids liable to undergo a strongly exothermic thermal decomposition even without participation of oxygen (air); other than materials classified as explosive, organic peroxides, or as oxidizing

Spontaneously combusting solids or pyrophoric solids are solids that, even in small quantities, are liable to ignite within 5 min after coming into contact with air. Substances and mixtures of this hazard class are assigned to a single hazard category on the basis of the outcome of the UN Test N.2. Self-heating substances are solids or liquids, other than a pyrophoric substance, which, by reaction with air and without energy supply, are liable to self-heat. Substances and mixtures of this hazard class are assigned to one of two hazard categories on the basis of the outcome of the UN Test N.4.

Substances which on contact with water emit flammable gases are liable to become spontaneously flammable or to give off flammable gases in dangerous quantities. Substances and mixtures of this hazard class are assigned to one of three hazard categories on the basis of the outcome of UN Test N.5, which measures gas evolution and speed of evolution. Flammable aerosols can be classified as Class 1 or Class 2 if they contain any component which is classified as flammable.

Oxidizing substances and organic peroxides contain

- Category 1: oxidizing substances
- Category 2: organic peroxides, organic liquids or solids that contain the bivalent -O-O- structure and may be considered derivatives of hydrogen peroxide, where one or both of the hydrogen atoms have been replaced by organic radicals. The term also includes organic peroxide formulations (mixtures).

Substances and mixtures of this hazard class are assigned to one of seven "Types," A to G, on the basis of the outcome of the UN Test Series A to H.

- Toxic and infectious substances
- Radioactive substances

Substances corrosive to metal are substances or mixtures that by chemical action will materially damage or even destroy metals. These substances or mixtures are classified in a single hazard category on the basis of tests (Steel: ISO 9328 (II): 1991—Steel type P235; Aluminum: ASTM G31-72 (1990)—non-clad types 7075-T6 or AZ5GU-T66). The GHS criteria are a corrosion rate on steel or aluminum surfaces exceeding 6.25 mm per year at a test temperature of 55°C.

Miscellaneous dangerous substances, which includes environmentally dangerous substances.

Health hazards

- Acute toxicity includes five GHS categories from which the appropriate elements relevant to transport, consumer, worker, and environmental protection can be selected. Substances are assigned to one of the five toxicity categories on the basis of LD50 (oral, dermal) or LC50 (inhalation).
- Skin corrosion means the production of irreversible damage to the skin following the application of a test substance for up to 4 h. Substances and mixtures in this hazard class are assigned to a single harmonized corrosion category.
- Skin irritation means the production of reversible damage to the skin following the application of a test substance for up to 4 h. Substances and mixtures in this hazard class are assigned to a single irritant category.
- For those authorities, such as pesticide regulators, wanting more than one designation for skin irritation, an additional mild irritant category is provided.
- Serious eye damage means the production of tissue damage in the eye, or serious physical decay of vision, following application of a test substance to the front surface of the eye, which is not fully reversible within 21 days of application. Substances and mixtures in this hazard class are assigned to a single harmonized category.
- Eye irritation means changes in the eye following the application of a test substance to the front surface of the eye, which are fully reversible within 21 days of application. Substances and mixtures in this hazard class are assigned to a single harmonized hazard category. For authorities, such as pesticide regulators, wanting more than one designation for eye irritation, one of two subcategories can be selected, depending on whether the effects are reversible in 21 or 7 days.

- Respiratory sensitizer means a substance that induces hypersensitivity of the airways following inhalation of the substance. Substances and mixtures in this hazard class are assigned to one hazard category.
- Skin sensitizer means a substance that will induce an allergic response following skin contact. The definition for "skin sensitizer" is equivalent to "contact sensitizer." Substances and mixtures in this hazard class are assigned to one hazard category.
- A germ cell mutagen is an agent giving rise to an increased occurrence of mutations in populations of cells and/or organisms. Substances and mixtures in this hazard class are assigned to one of two hazard categories. Category 1 has two subcategories.
- A carcinogen is a chemical substance or a mixture of chemical substances that induces cancer or increases its incidence. Substances and mixtures in this hazard class are assigned to one of two hazard categories. Category 1 has two subcategories.
- Reproductive toxins induce adverse effects on sexual function and fertility in adult males and females, as well as developmental toxicity in offspring. Substances and mixtures with reproductive and/or developmental effects are assigned to one of two hazard categories, "known or presumed" and "suspected." Category 1 has two subcategories for reproductive and developmental effects. Materials which cause concern for the health of breastfed children have a separate category: effects on or via lactation.
- The specific target organ toxicity (STOT) category distinguishes between single and repeated exposure for target organ effects. All significant health effects, not otherwise specifically included in the GHS that can impair function, both reversible and irreversible, immediate and/or delayed, are included in the non-lethal target organ/systemic toxicity class (TOST). Narcotic effects and respiratory tract irritation are considered to be target organ systemic effects following a single exposure. Substances and mixtures of the single exposure target organ toxicity hazard class are assigned to one of three hazard categories. Substances and mixtures of the repeated exposure target organ toxicity hazard class are assigned to one of two hazard categories.
- Aspiration hazard includes severe acute effects such as chemical pneumonia, varying degrees of pulmonary injury, or death following aspiration. Aspiration is the entry of a liquid or solid directly through the oral or nasal cavity, or indirectly from vomiting, into the trachea and lower respiratory system. Substances and mixtures of this hazard class are assigned to one of two hazard categories on the basis of viscosity.

Environmental hazards

- Acute aquatic toxicity means the intrinsic property of a material causing injury to an aquatic organism in a short-term exposure. Substances and mixtures of this hazard class are assigned to one of three toxicity categories on the basis of acute toxicity data: LC50 (fish), EC50 (crustacean), or ErC50 (for algae or other aquatic plants).
- In some regulatory systems these acute toxicity categories may be subdivided or extended for certain sectors.
- Chronic aquatic toxicity means the potential or actual properties of a material causing adverse effects to aquatic organisms during exposures that are determined in relation to the lifecycle of the organism. Substances and mixtures in this hazard class are assigned to one of four toxicity categories on the basis of acute data and environmental fate data: LC50 (fish), EC50 (crustacean), or ErC50 (for algae or other aquatic plants); and degradation or bioaccumulation.

Classification of mixtures

The GHS approach to the classification of mixtures for health and environmental hazards is also complex. It uses a tiered approach, and is dependent upon the amount of information available for the mixture itself and for its components. Principles have been developed for the classification of mixtures, drawing on existing systems, such as the European Union (EU) system for classification of preparations laid down in Directive 1999/45/EC. The process for the classification of mixtures is based on the following steps:

- Where toxicological or ecotoxicological test data are available for the mixture itself, the classification of the mixture will be based on that data.
- Where test data are not available for the mixture itself, then the appropriate bridging principles should be applied, which use test data for components and/or similar mixtures.
- If (1) test data are not available for the mixture itself, and (2) the bridging principles cannot be applied, then use the calculation or cutoff values described in the specific endpoint to classify the mixture.

Hazard communication

After the substance or mixture has been classified according to the GHS criteria, the hazards need to be communicated. As with many existing systems, the communication methods incorporated in GHS include labels and safety data sheets (SDSs). The GHS attempts to standardize hazard communication so that the intended audience can better understand the hazards of the chemicals in use. The GHS has established guiding principles.

The problem of trade secret or confidential business information has not been addressed within the GHS, except in general terms. For example, non-disclosure of confidential business information should not compromise the health and safety of users.

- Hazard communication should be available in more than one form (e.g., placards, labels or SDSs)
- Hazard communication should include hazard statements and precautionary statements.
- Hazard communication information should be easy to understand and standardized
- Hazard communication phrases should be consistent with each other to reduce confusion
- Hazard communication should take into account all existing research and any new evidence

Comprehensibility is challenging for a single culture and language. Global harmonization has numerous complexities. Some factors that affected the work include

- Different philosophies in existing systems on how and what should be communicated
- Language differences around the world
- Ability to translate phrases meaningfully
- Ability to understand and appropriately respond to symbols/ pictograms

GHS label elements

The symbol for substances hazardous to human health as implemented by the GHS.

The standardized label elements included in the GHS are

- Symbols (GHS hazard pictograms): Convey health, physical, and environmental hazard information, assigned to a GHS hazard class and category. Pictograms include the harmonized hazard symbols plus other graphic elements, such as borders, background patterns or cozers, and substances which have target organ toxicity.
- Also, harmful chemicals and irritants are marked with an exclamation mark, replacing the European saltire. Pictograms will have a black symbol on a white background with a red diamond frame. For transport, pictograms will have the background, symbol, and colors

currently used in the UN Recommendations on the Transport of Dangerous Goods. Where a transport pictogram appears, the GHS pictogram for the same hazard should not appear.

- Signal words: "Danger" or "warning" will be used to emphasize hazards and indicate the relative level of severity of the hazard, assigned to a GHS hazard class and category. Some lower-level hazard categories do not use signal words. Only one signal word corresponding to the class of the most severe hazard should be used on a label.
- Hazard statements: Standard phrases assigned to a hazard class and category that describe the nature of the hazard. An appropriate statement for each GHS hazard should be included on the label for products possessing more than one hazard.

The additional label elements included in the GHS are

- Precautionary statements: Measures to minimize or prevent adverse effects. There are four types of precautionary statements covering prevention, response in cases of accidental spillage or exposure, storage, and disposal. The precautionary statements have been linked to each GHS hazard statement and type of hazard.
- Product identifier (ingredient disclosure): Name or number used for a hazardous product on a label or in the SDS. The GHS label for a substance should include the chemical identity of the substance. For mixtures, the label should include the chemical identities of all ingredients that contribute to acute toxicity, skin corrosion, or serious eye damage, germ cell mutagenicity, carcinogenicity, reproductive toxicity, skin or respiratory sensitization, or target organ systemic toxicity (TOST), when these hazards appear on the label.
- Supplier identification: The name, address and telephone number should be provided on the label.
- Supplemental information: Non-harmonized information on the container of a hazardous product that is not required or specified under the GHS. Supplemental information may be used to provide further detail that does not contradict or cast doubt on the validity of the standardized hazard information.

GHS label format

The GHS includes directions for application of the hazard communication elements on the label. In particular, it specifies for each hazard, and for each class within the hazard, what signal word, pictogram, and hazard statement should be used. The GHS hazard pictograms, signal words, and hazard statements should be located together on the label.

The actual label format or layout is not specified in the GHS. National authorities may choose to specify where information should appear on the label or allow supplier discretion. There has been discussion about the size of GHS pictograms and that a GHS pictogram might be confused with a transport pictogram or "diamond."

Transport pictograms are different in appearance to the GHS pictograms. Annex 7 of the Purple Book explains how the GHS pictograms are expected to be proportional to the size of the label text, so that generally the GHS pictograms would be smaller than the transport pictograms.

GHS material safety data sheet or safety data sheet

The safety data sheet. The GHS has dropped the word "material" from "material safety data sheet." It will now be called the safety data sheet, or SDS. It is specifically aimed at use in the workplace. It should provide comprehensive information about the chemical product that allows employers and workers to obtain concise, relevant, and accurate information that can be put in perspective with regard to the hazards, uses, and risk management of the chemical product in the workplace. The SDS should contain 16 sections.

While there were some differences in existing industry recommendations, and requirements of countries, there was widespread agreement on a 16-section SDS that includes the following headings in the order specified:

1. Identification
2. Hazard(s) identification
3. Composition/information on ingredients
4. First-aid measures
5. Fire-fighting measures
6. Accidental release measures
7. Handling and storage
8. Exposure control/personal protection
9. Physical and chemical properties
10. Stability and reactivity
11. Toxicological information
12. Ecological information
13. Disposal considerations
14. Transport information
15. Regulatory information
16. Other information

The primary difference between the GHS requirements in terms of headings and sections and the international industry recommendations is that sections 2 and 3 have been reversed in order. The GHS SDS headings, sequence, and content are similar to the ISO, EU, and ANSI MSDS/SDS requirements.

The SDS should provide a clear description of the data used to identify the hazards. There is a table comparing the content and format of a current MSDS/SDS versus the GHS SDS provided in Appendix A of the OSHA GHS guidance document available.

Training

Current training procedures for hazard communication in the United States are more detailed than the GHS training recommendations. Therefore, educating employees on the updated chemical and product classifications and related pictograms, signal words, hazard statements, and precautionary measures will represent the greatest training challenge. Training will be a key component of the overall GHS approach, and should incorporate information as it is introduced into the workplace.

Employees and emergency responders will need to be trained on all new program elements, from hazard statements to pictograms. Bear in mind, if products are imported from countries that implement GHS prior to the United States and Canada, employee training may need to begin earlier than expected.

Implementation

The adoption of the GHS is expected to facilitate international trade, by increasing consistency between the laws in different countries that currently have different hazard communication requirements. There is no set international implementation schedule for the GHS. The goal of the United Nations was broad international adoption by 2008. Different countries will require different time frames to update current regulations or implement new ones.

Conclusion

In this handbook, I have attempted to include as much information as possible, in a format which can be easily understood by the first responder community and first responders. This handbook covers a variety of scenarios that have happened, and will continue to happen, in this country and other countries around the world. Today's world is a dangerous place, and we must prepare to handle and deal with potential attacks, not only from what we would call the typical terrorist from the Middle East, but our own home-grown terrorist groups within this country.

I do not claim to possess all of the expertise and answers, but have been a student, responder, and teacher within the responder community for the last 30 years. I have spent months in research and seeking information from many resources to come up with this overall, comprehensive handbook, and it is my hope that it will prove a useful resource to help and assist other first responders, educators, students, and everyday citizens.

As the twenty-first century unfolds, the threat of employment of weapons of mass destruction (WMD) against the U.S. homeland continues to be a reality. The new century was ushered in with the September 11, 2001 terrorist attacks; employment of the biological warfare agent anthrax; attacks against U.S. forces in Iraq using the chemical agent chlorine; attacks against airlines and in urban areas, such as Oklahoma City; and the Boston Bombing. Now, the death of Osama bin Laden could prompt revenge attacks, possibly using a radiological "dirty bomb" or some other WMD device, by terrorist cells. WMD threat planning and response remain high-priority endeavors for our state, local, and federal officials.

The purpose of this handbook remains twofold. It is an overall reference for the first responder who may be called to a site where WMD have been employed. It is also a primer for the average citizen desiring practical information on threat agents and procedures for surviving a terrorist attack involving WMD. Unlike other references, this handbook covers chemical, nuclear, radiological, biological, security management, threat assessment, school threat assessment, terror and terrorism, and explosives being used as weapons. The handbook presents straightforward, easily understood, and potentially life-saving information.

This handbook provides an integrated compilation of responder skills, knowledge, and capabilities. Each part provides the knowledge and the ability to train, and while perform, planning, and management are divided into specific response disciplines, there are many commonalities. These commonalities reflect the reality that effective WMD response strategies must be built on interoperability and an understanding of how all the pieces of the response fit together. This handbook helps illustrate the areas where common training and understanding, even cross-training among disciplines, can be effectively accomplished.

Always remember that this is not an exhaustive handbook for WMD incident-response abilities, but a general guide to help assist and guide the first responder community.

Glossary

Action plan See *incident action plan*.

Agency An agency is a division of government with a specific function, or a nongovernmental organization (e.g., private contractor, business, etc.) that offers a particular kind of assistance. In ICS, agencies are defined as jurisdictional (having statutory responsibility for incident mitigation) or assisting and/or cooperating (providing resources and/or assistance). (See *assisting agency*, *cooperating agency*, *jurisdictional agency*, and *multiagency incident*.)

Agency administrator or executive Chief executive officer (or designee) of the agency or jurisdiction that has responsibility for the incident.

Agency dispatch The agency or jurisdictional facility from which resources are allocated to incidents.

Agency representative An individual, assigned to an incident from an assisting or cooperating agency, who has been delegated authority to make decisions on matters affecting that agency's participation at the incident. Agency representatives report to the Incident Liaison Officer.

Air Operations Branch Director The person primarily responsible for preparing and implementing the air operations portion of the incident action plan. Also responsible for providing logistical support to helicopters operating at the incident.

Allocated resources Resources dispatched to an incident.

All-risk Any incident or event, natural or man-made, which warrants action to protect life, property, environment, and public health and safety, and minimize disruption of governmental, social, and economic activities.

Area Command An organization established to oversee the management of (1) multiple incidents that are each being handled by an ICS organization, or (2) large or multiple incidents to which several Incident Management Teams have been assigned. Area Command has a responsibility to set overall strategy and priorities, allocate critical resources according to priorities, ensure that incidents are properly managed, and ensure that objectives are met and strategies followed. Area Command becomes *Unified Area Command* when incidents are multijurisdictional. Area Command may be established at an emergency operations center facility or at some location other than an Incident Command Post.

Assigned resources Resources checked in and assigned work tasks on an incident.

Assignments Tasks given to resources to perform within a given operational period, based upon tactical objectives in the incident action plan.

Assistant Title for subordinates of the Command Staff positions. The title indicates a level of technical capability, qualifications, and responsibility subordinate to the primary positions.

Assisting agency An agency or organization providing personnel, services, or other resources to the agency with direct responsibility for incident management.

Available resources Resources assigned to an incident, checked in, and available for a mission assignment, normally located in a Staging Area.

Base The location at which primary logistics functions for an incident are coordinated and administered. There is only one base per incident. (Incident name or other designator will be added to the term *base*.) The Incident Command Post may be collocated with the base.

Branch The organizational level having functional or geographic responsibility for major parts of the operations or logistics functions. The Branch level is, organizationally, between Section and Division/Group in the Operations Section, and between Section and Units in the Logistics Section. Branches are identified by the use of Roman numerals or by functional name (e.g., medical, security, etc.).

Cache A pre-determined complement of tools, equipment, and/or supplies stored in a designated location, available for incident use.

Camp A geographical site, within the general incident area, separate from the incident base, equipped and staffed to provide sleeping, food, water, and sanitary services to incident personnel.

Chain of command A series of management positions in order of authority.

Check-in The process whereby resources first report to an incident. Check-in locations include: Incident Command Post (Resources Unit), incident base, camps, staging areas, helibases, helispots, and division supervisors (for direct line assignments).

Chief The ICS title for individuals responsible for functional sections: operations, planning, logistics, and finance/administration.

Clear text The use of plain English in radio communications transmissions. No 10 codes or agency-specific codes are used when utilizing clear text.

Command The act of directing and/or controlling resources by virtue of explicit legal, agency, or delegated authority. May also refer to the Incident Commander.

Command Post See *Incident Command Post*.

Command Staff The Command Staff consists of the Public Information Officer, Safety Officer, and Liaison Officer. They report directly to the Incident Commander. They may have an assistant or assistants, as needed.

Communications Unit An organizational unit in the Logistics Section responsible for providing communication services at an incident. A communications unit may also be a facility (e.g., a trailer or mobile van) used to provide the major part of an Incident Communications Center.

Compacts Formal working agreements among agencies to obtain mutual aid.

Compensation/Claims Unit Functional unit within the Finance/Administration Section responsible for financial concerns resulting from property damage, injuries, or fatalities at the incident.

Complex Two or more individual incidents located in the same general area assigned to a single Incident Commander or to Unified Command.

Cooperating agency An agency supplying assistance, other than direct operational or support functions, or resources to the incident management effort.

Coordination The process of systematically analyzing a situation, developing relevant information, and informing appropriate command authority of viable alternatives for selection of the most effective combination of available resources to meet specific objectives. The coordination process (which can be either intra- or inter-agency) does not involve dispatch actions. However, personnel responsible for coordination may perform command or dispatch functions within the limits established by specific agency delegations, procedures, legal authority, and so on.

Coordination center A facility that is used for the coordination of agency or jurisdictional resources in support of one or more incidents.

Cost sharing agreements Agreements between agencies or jurisdictions to share designated costs related to incidents. Cost sharing agreements are normally written but may also be oral between authorized agency and jurisdictional representatives at the incident.

Cost unit Functional unit within the Finance/Administration Section responsible for tracking costs, analyzing cost data, making cost estimates, and recommending cost-saving measures.

Crew See *single resource*.

Delegation of authority A statement provided to the Incident Commander by the agency executive delegating authority and assigning responsibility. The delegation of authority can include objectives, priorities, expectations, constraints, and other considerations or guidelines as needed. Many agencies require written delegation of authority to be given to Incident Commanders prior to their assuming command on larger incidents.

Demobilization unit Functional unit within the Planning Section responsible for ensuring orderly, safe, and efficient demobilization of incident resources.

Deputy A fully qualified individual who, in the absence of a superior, could be delegated the authority to manage a functional operation or perform a specific task. In some cases, a deputy could act as relief for a superior and therefore must be fully qualified in the position. Deputies can be assigned to the Incident Commander, General Staff, and Branch Directors.

Director The ICS title for individuals responsible for supervision of a Branch.

Dispatch The implementation of a command decision to move a resource or resources from one place to another.

Dispatch center A facility from which resources are ordered, mobilized, and assigned to an incident.

Division Divisions are used to divide an incident into geographical areas of operation. A Division is located within the ICS organization between the Branch and the Task Force/Strike Team. (See *group*.) Divisions are identified by alphabetic characters for horizontal applications and, often, by floor numbers when used in buildings.

Documentation unit Functional unit within the Planning Section responsible for collecting, recording, and safeguarding all documents relevant to the incident.

Emergency Absent a presidentially declared emergency, any incident(s), human-caused or natural, that requires responsive action to protect life or property. Under the Robert T. Stafford Disaster Relief and Emergency Assistance Act, an emergency means any occasion or instance for which, in the determination of the President, federal assistance is needed to supplement state and local efforts and capabilities to save lives and to protect property and public health and safety, or to lessen or avert the threat of a catastrophe in any part of the United States.

Emergency Management Coordinator/Director The individual within each political subdivision that has coordination responsibility for jurisdictional emergency management.

Emergency operations centers (EOCs) The physical locations at which the coordination of information and resources to support domestic incident management activities normally takes place. An EOC may be a temporary facility or may be located in a more central or permanently established facility, perhaps at a higher level of organization within a jurisdiction. EOCs may be organized by major functional disciplines (e.g., fire, law enforcement, and medical services), by jurisdiction (e.g., federal, state, regional, county, city, tribal), or some combination thereof.

Emergency operations plan (EOP) The plan that each jurisdiction has and maintains for responding to appropriate hazards.

Event A planned, non-emergency activity. ICS can be used as the management system for a wide range of events, for example, parades, concerts, or sporting events.

Facilities unit Functional unit within the Support Branch of the Logistics Section that provides fixed facilities for the incident. These facilities may include the incident base, feeding areas, sleeping areas, sanitary facilities, and so on.

Federal Of, or pertaining to, the federal government of the United States of America.

Field operations guide A pocket-sized manual of instructions on the application of the Incident Command System.

Finance/Administration Section The Section responsible for all incident costs and financial considerations. Includes the Time Unit, Procurement Unit, Compensation/Claims Unit, and Cost Unit.

Food unit Functional Unit within the Service Branch of the Logistics Section responsible for providing meals for incident personnel.

Function Function refers to the five major activities in ICS: command, operations, planning, logistics, and finance/administration. The term *function* is also used when describing the activity involved, for example, the planning function. A sixth function, intelligence, may be established, if required, to meet incident management needs.

General staff A group of incident management personnel organized according to function and reporting to the Incident Commander. The General Staff normally consists of the Operations Section Chief, Planning Section Chief, Logistics Section Chief, and Finance/Administration Section Chief.

Ground support unit Functional unit within the Support Branch of the Logistics Section responsible for the fueling, maintaining, and repairing of vehicles, and the transportation of personnel and supplies.

Group Groups are established to divide the incident into functional areas of operation. Groups are composed of resources assembled to perform a special function not necessarily within a single geographic division. Groups are located between Branches (when activated) and resources in the Operations Section.

Hazard Something that is potentially dangerous or harmful, often the root cause of an unwanted outcome.

Helibase The main location for parking, fueling, maintenance, and loading of helicopters operating in support of an incident. It is usually located at or near the incident base.

Helispot Any designated location where a helicopter can safely take off and land. Some helispots may be used for loading of supplies, equipment, or personnel.

Hierarchy of command See *chain of command*

Incident An occurrence or event, natural or man-made, which requires an emergency response to protect life or property. Incidents can, for example, include major disasters, emergencies, terrorist attacks, terrorist threats, wild land and urban fires, floods, hazardous materials spills, nuclear accidents, aircraft accidents, earthquakes, hurricanes, tornadoes, tropical storms, war-related disasters, public health and medical emergencies, and other occurrences requiring an emergency response.

Incident action plan (IAP) An oral or written plan containing general objectives reflecting the overall strategy for managing an incident. It may include the identification of operational resources and assignments. It may also include attachments that provide direction and important information for management of the incident during one or more operational periods.

Incident base Location at the incident where the primary logistics functions are coordinated and administered. (Incident name or other designator will be added to the term *base*.) The Incident Command Post may be collocated with the base. There is only one base per incident.

Incident Commander (IC) The individual responsible for all incident activities, including the development of strategies and tactics and the ordering and release of resources. The IC has overall authority and responsibility for conducting incident operations and is responsible for the management of all incident operations at the incident site.

Incident Command Post (ICP) The field location at which the primary tactical-level, on-scene incident command functions are performed. The ICP may be collocated with the incident base or other incident facilities and is normally identified by a green rotating or flashing light.

Incident Command System (ICS) A standardized on-scene emergency management construct, specifically designed to provide for the adoption of an integrated organizational structure that reflects the complexity and demands of single or multiple incidents, without being hindered by jurisdictional boundaries. ICS is the combination of facilities, equipment, personnel, procedures, and communications operating within a common organizational structure, designed to aid in the management of resources during incidents. It is used for all kinds of emergencies, and is applicable to small as well as large and complex incidents. ICS is used by various jurisdictions and functional agencies, both public and private, to organize field-level incident management operations.

Incident Communications Center The location of the Communications Unit and the Message Center.

Incident complex Two or more distinct incidents in the same general area that, by management action, are managed under a single incident commander or unified command in order to improve efficiency and simplify incident management processes.

Incident management team (IMT) The Incident Commander and appropriate Command and General Staff personnel assigned to an incident.

Incident objectives Statements of guidance and direction necessary for the selection of appropriate strategy and the tactical direction of resources. Incident objectives are based on realistic expectations of what can be accomplished when all allocated resources have been effectively deployed. Incident objectives must be achievable and measurable, yet flexible enough to allow for strategic and tactical alternatives.

Incident support organization Includes any off-incident support provided to an incident. Examples would include agency dispatch centers, airports, mobilization centers, and so on.

Incident types Incidents are categorized by five types based on complexity. Type 5 incidents are the least complex and Type 1 the most complex.

Initial action The actions taken by resources that are the first to arrive at an incident site.

Initial response Resources initially committed to an incident.

Intelligence officer The Intelligence Officer is responsible for managing internal information, intelligence, and operational security requirements supporting incident management activities. These may include information security and operational security activities, as well as the complex task of ensuring that sensitive information of all types (e.g., classified information, law enforcement sensitive information, proprietary information, or export-controlled information) is handled in a way that not only safeguards the information, but also ensures that it gets to those who need access to it to perform their missions effectively and safely.

Joint Information Center (JIC) A facility established to coordinate all incident-related public information activities. It is the central point of contact for all news media at the scene of the incident. Public information officials from all participating agencies should collocate at the JIC.

Joint information system (JIS) Integrates incident information and public affairs into a cohesive organization designed to provide consistent, coordinated, timely information during crisis or incident operations. The mission of the JIS is to provide a structure and system for developing and delivering coordinated interagency messages; developing, recommending, and executing public information plans and strategies on behalf of the Incident Commander; advising the Incident Commander concerning public affairs issues that could affect a response effort; and controlling rumors and inaccurate information that could undermine public confidence in the emergency response effort.

Jurisdiction A range or sphere of authority. Public agencies have jurisdiction at an incident related to their legal responsibilities and authority. Jurisdictional authority at an incident can be political or geographical (e.g., city, county, tribal, state, or federal boundary lines) or functional (e.g., law enforcement, public health).

Jurisdictional agency The agency having jurisdiction and responsibility for a specific geographical area, or a mandated function.

Kinds of resources Describe what the resource is (e.g., medic, firefighter, Planning Section Chief, helicopters, ambulances, combustible gas indicators, bulldozers).

Landing zone See *helispot*.

Leader The ICS title for an individual responsible for a task force, strike team, or functional unit.

Liaison A form of communication for establishing and maintaining mutual understanding and cooperation.

Liaison officer (LNO) A member of the Command Staff responsible for coordinating with representatives from cooperating and assisting agencies. The Liaison Officer may have assistants.

Logistics Providing resources and other services to support incident management

Logistics section The section responsible for providing facilities, services, and materials for the incident.

Local government A county, municipality, city, town, township, local public authority, school district, special district, intrastate district, council of governments (regardless of whether the council of governments is incorporated as a non-profit corporation under state law), regional or interstate government entity, or agency or instrumentality of a local government; an Indian tribe or authorized tribal organization, or in Alaska, a native village or Alaska Regional Native Corporation; a rural community, unincorporated town or village, or other public entity. See Section 2 (10), Homeland Security Act of 2002, Public Law 107-296, 116 Stat. 2135 (2002).

Major disaster As defined under the Robert T. Stafford Disaster Relief and Emergency Assistance Act (42 U.S.C. 5122), a major disaster is any natural catastrophe (including any hurricane, tornado, storm, high water, wind-driven water, tidal wave, tsunami, earthquake, volcanic eruption, landslide, mudslide, snowstorm, or drought), or, regardless of cause, any fire, flood, or explosion, in any part of the United States, which in the determination of the President causes damage of sufficient severity and magnitude to warrant major disaster assistance under this Act to supplement the efforts and available resources of states, tribes, local governments, and disaster relief organizations in alleviating the damage, loss, hardship, or suffering caused thereby.

Management by objective A management approach that involves a four-step process for achieving the incident goal. The management-by-objectives approach includes the following: establishing overarching objectives; developing and issuing assignments, plans, procedures, and protocols; establishing specific, measurable objectives for various incident management functional activities and directing efforts to fulfill them, in support of defined strategic objectives; and documenting results to measure performance and facilitate corrective action.

Managers Individuals within ICS organizational units that are assigned specific managerial responsibilities, for example, Staging Area Manager or Camp Manager.

Medical unit Functional Unit within the Service Branch of the Logistics Section responsible for the development of the medical emergency plan, and for providing emergency medical treatment of incident personnel.

Message center The Message Center is part of the Incident Communications Center and is collocated or placed adjacent to it. It receives records, and routes information about resources reporting to the incident, resource status, and administrative and tactical traffic.

Mitigation The activities designed to reduce or eliminate risks to persons or property or to lessen the actual or potential effects or consequences of an incident. Mitigation measures may be implemented prior to, during, or after an incident. Mitigation measures are often formed by lessons learned from prior incidents. Mitigation involves ongoing actions to reduce exposure to, probability of, or potential loss from, hazards. Measures may include zoning and building codes, floodplain buyouts, and analysis of hazard-related data to determine where it is safe to build or locate temporary facilities. Mitigation can include efforts to educate governments, businesses, and the public on measures they can take to reduce loss and injury.

Mobilization The process and procedures used by all organizations (federal, state, and local) for activating, assembling, and transporting all resources that have been requested to respond to or support an incident.

Mobilization center An off-incident location at which emergency service personnel and equipment are temporarily located pending assignment, release, or reassignment.

Multiagency coordination (MAC) The coordination of assisting agency resources and support to emergency operations.

Multiagency coordination systems (MACS) Multiagency coordination systems provide the architecture to support coordination for incident prioritization, critical resource allocation, communications systems integration, and information coordination. The components of multiagency coordination systems include facilities, equipment, emergency operations centers (EOCs), specific multiagency coordination entities, personnel, procedures, and communications. These systems assist agencies and organizations to fully integrate the subsystems of the NIMS.

Multiagency incident An incident where one or more agencies assist a jurisdictional agency or agencies. May be single or unified command.

Mutual-aid agreement Written agreement between agencies and/or jurisdictions that they will assist one another on request, by furnishing personnel, equipment, and/or expertise in a specified manner.

National Incident Management System (NIMS) A system mandated by HSPD-5 that provides a consistent nationwide approach for federal, state, local, and tribal governments; the private sector; and nongovernmental organizations to work effectively and efficiently together to prepare for, respond to, and recover from, domestic incidents, regardless of cause, size, or complexity. To provide for interoperability and compatibility among federal, state, local, and tribal capabilities, the NIMS includes a core set of concepts, principles, and terminology. HSPD-5 identifies these as the ICS; multiagency coordination systems; training; identification and management of resources (including systems for classifying types of resources); qualification and certification; and the collection, tracking, and reporting of incident information and incident resources.

Officer The ICS title for the personnel responsible for the Command Staff positions of Safety, Liaison, and Public Information.

Operational period The period of time scheduled for execution of a given set of operation actions as specified in the incident action plan.

Operational periods can be of various lengths, although usually not over 24 h.

Operations section The section responsible for all tactical operations at the incident. Includes branches, divisions and/or groups, task forces, strike teams, single resources, and staging areas.

Out-of-service resources Resources assigned to an incident but unable to respond for mechanical, rest, or personnel reasons.

Planning meeting A meeting held as needed throughout the duration of an incident, to select specific strategies and tactics for incident control operations, and for service and support planning. On larger incidents, the planning meeting is a major element in the development of the incident action plan.

Planning section Responsible for the collection, evaluation, and dissemination of information related to the incident, and for the preparation and documentation of incident action plans. The Section also maintains information on the current and forecasted situation, and on the status of resources assigned to the incident. Includes the Situation, Resources, Documentation, and Demobilization Units, as well as technical specialists.

Preparedness The range of deliberate, critical tasks and activities necessary to build, sustain, and improve the operational capability to prevent, protect against, respond to, and recover from, domestic incidents. Preparedness is a continuous process. Preparedness involves efforts at all levels of government and between government, private-sector, and nongovernmental organizations to identify threats, determine vulnerabilities, and identify required resources. Within the NIMS, preparedness is operationally focused on establishing guidelines, protocols, and standards for planning, training and exercises, personnel qualification and certification, equipment certification, and publication management.

Preparedness organizations The groups that provide interagency coordination for domestic incident management activities in a nonemergency context. Preparedness organizations can include all agencies with a role in incident management for prevention, preparedness, response, or recovery activities. They represent a wide variety of committees, planning groups, and other organizations that meet and coordinate to ensure the proper level of planning, training, equipping, and other preparedness requirements within a jurisdiction or area.

Prevention Actions to avoid an incident or to intervene to stop an incident from occurring. Prevention involves actions to protect lives and property. It involves applying intelligence and other information to a range of activities that may include such countermeasures as deterrence operations; heightened inspections; improved surveillance and security operations; investigations to determine the full nature and source of the threat; public health and agricultural surveillance and testing processes; immunizations, isolation, or quarantine; and, as appropriate, specific law enforcement operations aimed at deterring, preempting, interdicting, or disrupting illegal activity and apprehending potential perpetrators and bringing them to justice.

Procurement unit Functional unit within the Finance/Administration Section responsible for financial matters involving vendor contracts.

Public Information Officer (PIO) A member of the Command Staff responsible for interfacing with the public and media or with other agencies with incident-related information requirements.

Recognition primed decision-making A model that describes how experts make decisions under stressful situations that are time critical and rapidly changing.

Recorders Individuals within ICS organizational units who are responsible for recording information. Recorders may be found in Planning, Logistics, and Finance/Administration Units.

Recovery The development, coordination, and execution of service- and site-restoration plans; the reconstitution of government operations and services; individual, private-sector, nongovernmental, and public-assistance programs to provide housing and to promote restoration; long-term care and treatment of affected persons; additional measures for social, political, environmental, and economic restoration; evaluation of the incident to identify lessons learned; post-incident reporting; and development of initiatives to mitigate the effects of future incidents.

Reinforced response Those resources requested in addition to the initial response.

Reporting locations Location or facilities where incoming resources can check-in at the incident.

Resource management Efficient incident management requires a system for identifying available resources at all jurisdictional levels to enable timely and unimpeded access to resources needed to prepare for, respond to, or recover from, an incident. Resource management under the NIMS includes mutual-aid agreements; the use of special federal, state, local, and tribal teams; and resource mobilization protocols.

Resources Personnel and major items of equipment, supplies, and facilities available or potentially available for assignment to incident operations and for which status is maintained. Resources are described by type, and may be used in operational support or supervisory capacities at an incident or at an EOC.

Resources Unit Functional Unit within the Planning Section responsible for recording the status of resources committed to the incident. The Unit also evaluates resources currently committed to the incident, the impact that additional responding resources will have on the incident, and anticipated resource needs.

Response Activities that address the short-term, direct effects of an incident. Response includes immediate actions to save lives, protect property, and meet basic human needs. Response also includes the execution of emergency operations plans and of mitigation activities designed to limit the loss of life, personal injury, property damage, and other unfavorable outcomes. As indicated by the situation, response activities include applying intelligence and other information to lessen the effects or consequences of an incident; increased security operations; continuing investigations into the nature and source of the threat; ongoing public health and agricultural surveillance and testing processes; immunizations, isolation, or quarantine; and specific law enforcement operations aimed at preempting, interdicting, or disrupting illegal activity, and apprehending actual perpetrators and bringing them to justice.

Safety Officer A member of the Command Staff responsible for monitoring and assessing safety hazards or unsafe situations, and for developing measures for ensuring personnel safety. The Safety Officer may have assistants.

Section The organizational level having responsibility for a major functional area of incident management, for example, operations, planning, logistics, finance/administration, and intelligence (if established). The Section is organizationally situated between the Branch and the Incident Command.

Segment A geographical area in which a task force/strike team leader or supervisor of a single resource is assigned authority and responsibility for the coordination of resources and implementation of planned tactics. A segment may be a portion of a division or an area inside or outside the perimeter of an incident. Segments are identified with Arabic numbers.

Service branch A branch within the Logistics Section responsible for service activities at the incident. Includes the Communication, Medical, and Food Units.

Single resource An individual, a piece of equipment and its personnel complement, or a crew or team of individuals with an identified work supervisor that can be used on an incident.

Situation unit Functional unit within the Planning Section responsible for the collection, organization, and analysis of incident status information, and for analysis of the situation as it progresses. Reports to the Planning Section Chief.

Span of control The number of individuals a supervisor is responsible for, usually expressed as the ratio of supervisors to individuals. (Under the NIMS, an appropriate span of control is between 1:3 and 1:7.)

Staging area Location established where resources can be placed while awaiting a tactical assignment. The Operations Section manages staging areas.

Standard operating procedure (SOP) Complete reference document or an operations manual that provides the purpose, authorities, duration, and details for the preferred method of performing a single function or a number of interrelated functions in a uniform manner.

State When capitalized, refers to a specific state of the United States, the District of Columbia, the Commonwealth of Puerto Rico, the Virgin Islands, Guam, American Samoa, the Commonwealth of the Northern Mariana Islands, and any possession of the United States. See Section 2 (14), Homeland Security Act of 2002, Public Law 107-296, 116 Stat. 2135 (2002).

Strategic Strategic elements of incident management are characterized by continuous long-term, high-level planning by organizations headed by elected or other senior officials. These elements involve the adoption of long-range goals and objectives, the setting of priorities, the establishment of budgets and other fiscal decisions, policy development, and the application of measures of performance or effectiveness.

Strategy The general direction selected to accomplish incident objectives set by the Incident Commander.

Strike team A specified combination of the same kind and type of resources with common communications and a leader.

Supervisor The ICS title for individuals responsible for a division or group.

Supply unit Functional unit within the Support Branch of the Logistics Section responsible for ordering equipment and supplies required for incident operations.

Support branch A branch within the Logistics Section responsible for providing personnel, equipment, and supplies to support incident operations. Includes the Supply, Facilities, and Ground Support Units.

Support resources Non tactical resources under the supervision of the Logistics, Planning, or Finance/Administration Sections or the Command Staff.

Supporting materials Refers to the several attachments that may be included with an incident action plan, for example, communications plan, map, safety plan, traffic plan, and medical plan.

Tactical direction Direction given by the Operations Section Chief that includes the tactics required to implement the selected strategy, the selection and assignment of resources to carry out the tactics, directions for tactic implementation, and performance monitoring for each operational period.

Tactics Deploying and directing resources on an incident to accomplish incident strategy and objectives.

Task force A combination of single resources assembled for a particular tactical need with common communications and a leader.

Team See *single resource*.

Technical specialists Personnel with special skills that can be used anywhere within the ICS organization.

Threat An indication of possible violence, harm, or danger.

Time Unit Functional unit within the Finance/Administration Section responsible for recording time for incident personnel and hired equipment.

Tools Those instruments and capabilities that allow for the professional performance of tasks, such as information systems, agreements, doctrine, capabilities, and legislative authorities.

Tribal Any Indian tribe, band, nation, or other organized group or community, including any Alaskan native village as defined in or established pursuant to the Alaskan Native Claims Settlement Act (85 Stat. 688) (43 U.S.C.A. and 1601 *et seq*.), that is recognized as eligible for the special programs and services provided by the United States to Indians because of their status as Indians.

Type A classification of resources in the ICS that refers to capability. Type 1 is generally considered to be more capable than Types 2, 3, or 4, because of size, power, capacity, or, in the case of Incident Management Teams, experience and qualifications

Unified Area Command A Unified Area Command is established when incidents under an Area Command are multijurisdictional. (See *Area Command* and *Unified Command*.)

Unified Command An application of ICS used when there is more than one agency with incident jurisdiction or when incidents cross political jurisdictions. Agencies work together through designated members of the Unified Command, often senior personnel from agencies and/or disciplines participating in the Unified Command, to establish a common set of objectives and strategies and a single incident action plan.

Unit The organizational element having functional responsibility for a specific incident planning, logistics, or finance/administration activity.

Unity of command The concept by which each person within an organization reports to one, and only one, designated person. The purpose of unity of command is to ensure unity of effort under one responsible commander for every objective.

Bibliography

Air Gas Inc. (2006). *Emergency Response Organizations (Trained. Equipped. Ready.)* Chicago, IL: AirGas.

Andersen, S. (2009). *Threats and Countermeasures.* Redmond, OR: Microsoft Corporation.

ASIS International. (2014). *ASIS International*. Retrieved May 22, 2014, from https://www.asisonline.org/Pages/default.aspx

Associates, A. (2011). *Director of Global Security*. Retrieved July 3, 2012, from http://www.devex.com/en/jobs/director-of-global-security

Atlas, R. I. (2008). *21st Century Security and CPTED*. Boca Raton, FL: Auerbach Publications.

Barenblatt, D. (2004). *A Plague upon Humanity*. London: HarperCollins.

BBC News. (2007). *China spying "biggest US threat"*. Retrieved May 22, 2012, from http://news.bbc.co.uk/2/hi/americas/7097296.stm

Benson, C. (2012). *Security strategies*. Retrieved June 29, 2012, from http://technet.microsoft.com/en-us/library/cc723506.aspx

Bernstein, J. (2012). *All about crisis management*. Retrieved July 24, 2012, from http://managementhelp.org/crisismanagement/index.htm

Bolin, A. (2007). HPT in Military Settings. *Military Review*, 46 (3), pp. 5–7.

Campbell, K. (2001). *To Prevail: An American Strategy for the Campaign against Terrorism*. Washington, DC: CSIS Press.

Central Intelligence Agency. (2007). *CIA & the war on terrorism*. Retrieved April 9, 2007, from https://www.cia.gov/news-information/cia-the-war-on-terrorism/index.html

Chapman, R. (2002). *COPS Innovations: A Closer Look: Law Enforcement Responds to Terrorism*. Washington, DC: U.S. Department of Justice.

Committee on Armed Services, United States Senate. (2008). *Subcommittee on emerging threats and capabilities*. Retrieved March 12, 2008, from http://www.gpo.gov/fdsys/pkg/CHRG-110shrg45110/html/CHRG-110shrg45110.htm

Deutch, J. (2002). Smarter Intelligence. *Foreign Policy*, no. 128, pp. 64–69.

Eggen, D. (2003). "Bush aims to blend counterterrorism efforts." February 15, 2003, *Washington Post*.

Emergency Management Institute. (2014). *National Incident Management System.* Retrieved August 29, 2014, from http://www.training.fema.gov/IS/NIMS.aspx

Federal Bureau of Investigation. (2014, August 29). *Weapons of mass destruction*. Retrieved August 29, 2014, from http://www.fbi.gov/about-us/investigate/terrorism/wmd/wmd

Federation of American Scientists. (2014). *Types of chemical weapons*. Retrieved August 29, 2014, from http://fas.org/programs/bio/chemweapons/cwagents.html

GlobalSecurity.org. (2011). *Terrorism information awareness*. Retrieved May 30, 2012, from http://www.globalsecurity.org/security/systems/tia.htm

GlobalSecurity.org. (2014). *Weapons of mass destruction (WMD)*. Retrieved August 29, 2014, from http://www.globalsecurity.org/wmd/index.html

Government of the United States. (2001). *The Computer Fraud and Abuse Act (as amended 1994 and 1996)*. Retrieved October 23, 2012, from http://www.panix.com/~eck/computer-fraud-act.html

Grabosky, P. (1998). *Crime and Technology in the Global Village*. Melbourne, Australia: Australian Institute of Criminology.

Hana, U. (2011). Staff turnover as a possible threat to knowledge loss. *Journal of Competitiveness*, 3, p. 84–98.

Henry, C. M. (2005). *The Clash of Globalisations in the Middle East*. Oxford: The University Press.

Homeland Security. (2012). *Aviation security policy*. Retrieved April 9, 2012, from www.dhs.gov/files/laws/gc_1173113497603.shtm

Homeland Security. (2012). *Chronology of changes to the homeland security advisory system*. Retrieved August 13, 2012, from http://www.dhs.gov/homeland-security-advisory-system

Homeland Security. (2014). *School safety*. Retrieved August 29, 2014, from http://www.dhs.gov/school-safety

Homeland Security News. (2014). *Dirty bomb*. Retrieved August 29, 2014, from http://www.nationalterroralert.com/dirtybomb/

Honeywell Airport Solutions. (2007, July 26). *Integrated Security Systems*. Bracknell, UK: Honeywell Corporation.

Hooker, E. (2014). *Biological Warfare*. Retrieved August 29, 2014, from http://www.emedicinehealth.com/biological_warfare/article_em.htm

Intergration, O. O. (2008). *Risk Managment*. Los Angeles, CA: State of California.

International Society of Explosives Engineers. (2014). *The world of explosives*. Retrieved September 2, 2014, from http://explosives.org/ and https://www.isee.org/

Kabay, M. E. (2007). *Industrial espionage*. Northfield, MN: CTO & MSIA Program Director.

Kraynak, B. (2002). *Hypothesizing the Cyber-Terrorist*. Washington, DC: George Washington University.

Kruse, W. G. (2002). *Computer Forensics: Incident Response Essentials*. Boston, MA: Addison-Wesley.

Langelaan, J. Y. (2012). *Industrial espionage*. Retrieved June 29, 2012, from http://html.rincondelvago.com/industrial-espionage.html

Leigland, R. (2004). Formalization of Digital Forensics. *International Journal of Digital Evidence*, pp. 1–32.

Management, M. F. (2008). *China: Opportunities and Risks for Foreign Companies*. London: Insight Investment Management.

Massa, D. (2007). *IAFF training for hazardous materials*. Retrieved December 18, 2012, from https://tools.niehs.nih.gov/wetp/public/Course_download2.cfm?tranid=4653. Recognition and identification: Diane Massa.

McGee, D. E. (2001). "FBI rushes to remake its mission: Counterterrorism focus replaces crime solving." November 12, 2001, *Washington Post*.

McGowan, T. (2013). *Hazardous Materials/Weapons of Mass Destruction Response Handbook NFPA 742 & 473*. Qunicy, MA: National Fire Protection Association.

Mendell, R. L. (2003). The quiet threat. In R. L. Mendell, *The Quiet Threat* (pp. 68–71). Springfield, IL: Charles C. Thomas.

Meyer, R. (2007). *Explosives*, 6th edn. Weinheim, Germany: Wiley.

Minieri, M. W. (2004). *Protecting Corporate Secrets*. Reston, VA: Kroll.

Moteff, J. (2005). *Risk Management and Critical Infrastructure*. Washington, DC: Congressional Research Service.

Nasheri, H. (2009). *Economic Espionage and Industrial Spying*. New York: Cambridge.

Northeast States Emergency Consortium. (2012). What is the risk of terrorism in the Northeast? Retrieved September 18, 2012, from http://www.nesec.org/hazards/terrorism.cfm

Nuclear Weapon Archive. (2014, August 29). *The nuclear weapon archive*. Retrieved August 29, 2014, from http://nuclearweaponarchive.org/

Organisation for the Prohibition of Chemical Weapons. (2014). *Brief description of chemical weapons*. Retrieved August 29, 2014, from http://www.opcw.org/about-chemical-weapons/what-is-a-chemical-weapon

Porterfield, W. W. (1993). *Inorganic Chemistry: A Unified Approach*, 2nd edn. San Diego, CA: Academic Press.

Price, M. (2008). The China security threat. *Information Age*.

Renfroe, N. A. (2014). *Threat/vulnerability assessments and risk analysis*. Retrieved August 18, 2014, from http://www.wbdg.org/resources/riskanalysis.php

Risk Protection Group (2012). *Personnel protection: Your employees are your most valuable asset*. Retrieved August 13, 2012, from http://riskprotection.us/id65.html

Sarathy, R. (2006). Security and the global supply chain. Retrieved July 31, 2012, from http://www.thefreelibrary.com/Security+and+the+global+supply+chain.-a0154866857

Security Management. (2014, June 12). *Security Management*. Retrieved June 12, 2014, from http://securitymanagement.com/

Shelley, L. (2012). *Terrorism, Transnational Crime and Corruption Center*. Retrieved July 24, 2012, from http://traccc.gmu.edu/

Shinder, D. L. (2007). *10 physical security measures every organization should take*. Retrieved June 29, 2012, from http://www.techrepublic.com/blog/10things/10-physical-security-measures-every-organization-should-take/106

Terrorism-Research.com. (2014). *What is terrorism?* Retrieved August 29, 2014, from http://www.terrorism-research.com/

United States of America. (1948). *Title 18: Crimes and Criminal Procedure Conspiracy to Commit Offense Or to Defraud United States*. Washington, DC.

Unites States-China Security Review Commission. (2002) Report Offers Recommendations on U.S.-China Relations (U.S.-China Security Review Commission 2002 Annual Report). Washington, DC: U.S. Department of State. Retrieved from https://www.ait.org.tw/en/20020715-report-offers-recommendations-on-us-relations.html

United States Congress, Office of Technology Assessment. (1993). *Technologies Underlying Weapons of Mass Destruction*. Washington, DC: Government Printing Office.

United States Department of Justice. (2014). *Bureau of Alcohol, Tobacco, Firearms and Explosives*. Retrieved August 19, 2014, from http://www.atf.gov/content/Explosives/explosives-industry?qt-explosive_tab=3#qt-explosive_tab

United Federation of Teachers. (2013). *School safety plans*. Retrieved September 26, 2013 from http://www.uft.org/know-your-rights/school-safety-plans

United Nations. (2014). *United Nations action to counter terrorism*. Retrieved August 29, 2014, from http://www.un.org/en/terrorism/

United States Army. (2005). *Potential Military Chemical/Biological Agents and Compounds. FM 3-11.9*. Washington, DC: GSA.

United States Secret Service. (2014). *National Threat Assessment Center*. Retrieved August 29, 2014, from http://www.secretservice.gov/ntac.shtml

United States Agency for International Development. (2010). *Cyber crime: Its impact on government, society and the prosecutor*. (2010). Retrieved September 11, 2012, from pdf.usaid.gov/pdf_docs/PNADA641.pdf

United States Department of State. (2003). *Coordinator for Counterterrorism.* Retrieved February 22, 2003, from http://www.state.gov/s/ct

United States Department of Defense. (2014). *Strategy for Countering Weapons of Mass Destruction*. Washington, DC: Department of Defense.

United States Department of Energy. (2012). *Espionage and foreign travel*. Retrieved June 29, 2012, from http://www.ch.doe.gov/offices/OCI/EspionageTravel/index.htm

United States Government Accountability Office. (2005) *Overseas security*. Retrieved July 3, 2012, from http://www.gao.gov/products/GAO-05-642

University of Pittsburgh. (2003, May 1). *Counterterrorism Policy.* Retrieved January 12, 2017, from http://jurist.law.pitt.edu/terrorism/terrorism2.htm

Wirtz, E. A. (2005). *Weapons of Mass Destruction.* Santa Barbara, CA: ABC-CLIO.

Zimmerman, P. D. (2004). Dirty bombs: The threat revisited. *Center for Technology and National Security Policy National Defense University*.

Index

9/11 Commission, *see* National Commission on Terrorist Attacks upon the United States
1794 Whiskey Rebellion, 51
1967 Detroit riot, 51–52
2005 United Nations International Convention for the Suppression of Acts of Nuclear Terrorism, 168
2011 DSCA Interagency Partner Guide, 50

Abel, Frederick Augustus, 156
Access control, 14
Accountability, 8
Acoustical eavesdropping, 41
Acute aquatic toxicity, 240
Acute radiation syndrome, 72
Acute toxicity, 238
Aesop, 14
Airborne contamination, 42
AIS, *see* Automated information system (AIS)
Alarm annunciation, 17, 18
Alarm clocks, 13
Alarm devices, 13, 13–14
Alarm management, 17–21
- concepts, 17–18
- improvement methods, 19–20
  - advanced, 20
  - design guide, 19
  - documentation and rationalization, 19
  - nuisance reduction, 19
- need for, 19
- steps to, 20–21
  - audit and enforcement, 21
  - control and maintain performance, 21
  - documentation and rationalization (D&R), 20
  - performance benchmarking, 20
  - philosophy, 20
  - real-time, 21
  - resolution, 20

Alexander, Keith B., 81, 83
Allison, Graham, 168
al-Qaeda, 115, 170
Ambulatory casualties, 198
American deployments, 53
Anthrax, 68–70
- cutaneous, 69
- gastrointestinal, 69
- inhalational, 69
- overview, 68–69
- vaccine, 69–70

Antimatter, 165
Antitank weapons, and mortars, 42
Anti-vehicle barriers, 15
Area Command, 5
Arms control and peacekeeping, 52–54
Ashcroft, John, xxviii
Aspiration hazard, 239
Assassins, *see* Hashhashin
ASTM Standard
- Guide F1001-86, 218
- Practice F1052, 217
- Test Method F903-86, 218

Astronomical explosion, 155
Attack dogs, 15
ATSA, *see* Aviation and Transportation Security Act (ATSA)
Aum Shinrikyo, 170
Australian deployments, 54
Autodialed alarm, 13
Automated information system (AIS), 27
Aviation and Transportation Security Act (ATSA), xxvi
Aviation security, 30

Bacchiocchi, Samuele, 113
The Balance of Terror: Strategy for the Nuclear Age, 167
Ballistic attack, 41
Barot, Dhiren, 178, 179
Barricade, 15
Bath, J., 120
Belfer Center for Science and International Affairs, 169, 170
Bike rack barricades, 15
bin Laden, Osama, 169, 170
Biological agents, 42
Biological weapons (BW), 126–128
  delivery systems, 128
  pathogen feed stocks, 127
Biological weapons of mass destruction, 65–71
  anthrax, 68–70
    cutaneous, 69
    gastrointestinal, 69
    inhalational, 69
    overview, 68–69
    vaccine, 69–70
  attractive to terrorists, 65
  bioterrorism, 65
  characteristics of attacks, 65
  critical biological agents, 66–67
  genealogical classification, 65–66
  plague, 70–71
    bubonic, 70
    overview, 70
    pneumonic, 70
    septicemic, 70–71
  smallpox, 67–68
    vs. chickenpox, 68
    clinical features, 68
    history, 67
    vaccine, 68
    Variola major, 67–68
Biometrics, 14
Bioterrorism, 65
Blast barriers, 15
Blasting agents, *see* Tertiary explosive
BLEVE, *see* Boiling liquid expanding vapor explosion (BLEVE)
Blister agents, 78
Blood agents, 64, 77–78
Body water loss, 225
Boiling liquid expanding vapor explosion (BLEVE), 233–236
  fires, 235–236
  mechanism, 233–234
  water example, 234–235
  without chemical reactions, 235
Boosted fission weapon, 164
The Boy Who Cried Wolf, 14
Brisance, 142–143, 154
British deployments, 53
Bubonic plague, 70
Burglar alarms, 13
Bush, George W., xxvi, 1
Business impact, 25–27
BW, *see* Biological weapons (BW)

C8-Direct Application Decontamination System (DADS), 191
Cancer, 177
Car alarms, 13
Carcinogen, 239
Casey, William, xxviii
Casualty processing, 200–201
CBRNE, *see* Chemical, biological, radiological, nuclear and explosive (CBRNE)
CBRNE enhanced response force package (CERFP), 106
CERCLA, *see* Comprehensive Environmental Response, Compensation, and Liability Act (CERCLA)
CERFP, *see* CBRNE enhanced response force package (CERFP)
Certified Information Systems Auditor Review Manual 2006, 25
Chain of command, 7
Chechen extremists, 178
Chemical, biological, radiological, nuclear and explosive (CBRNE), and WMD, 123–135
  biological, 126–128
    BW delivery systems, 128
    pathogen feed stocks, 127
  chemical, 123–125
  industrial chemicals, 131–135
    agents and terrorism, 134
    chlorine and anhydrous ammonia, 133–134
    indicators of chemical incident, 134–135
    pesticides and rodenticides, 132–133
    tactics, 131–132
  nuclear, 129–131
  radiological, 128–129
Chemical agents, 42

Chemical constitution, 140–141
Chemical decomposition, 138
Chemical explosives, 156
Chemical hazards, 207
Chemical protective clothing, 181, 205–226
applications, 205–207
classification of, 209–213
decontamination for, reuse, 185
donning and doffing, 218–221
inspection, 222
maintenance, 223
storage, 222–223
training, 224
user monitoring, 221
level of, 207
risks, 224–225
body water loss, 225
heat stress, 224–225
selection factors, 214–218
ensemble, 207–209
general guidelines, 216–218
physical properties, 215–216
sources of information, 215
service life of, 213–214
Chemical resistance, material, 214
Chemical victims, types of, 197–201
casualty processing, 200–201
decontamination prioritization, 198–200
mass, processing tag system, 200
triage, 198–200
Chemical warfare (CW) agents, 77–80
blister agents, 78
blood agents, 77–78
choking agents, 77
control zones, 79–80
isolate, 79
lethality relative to chlorine, 78–79
nerve agents, 78
Chemical warfare, and industrial agents, 187–203
decontaminants, 187–188
decontamination approaches, 196–197
field-expedient water methods, 196–197
non-aqueous methods, 197
decontamination methods, 188–192
equipment, 190–192
individual, 189–190
environmental concerns, 201
guidelines for mass casualty decontamination, 192–196
methods, 194–195
procedures, 195–196
purposes, 193–194
relationships and process, 192–193
overview, 187
types of chemical victims, 197–201
casualty processing, 200–201
prioritization, 198–200
Chemical weapons (CW), 123–125
Chemical weapons (CW) of mass destruction, 62–65
classification, 62–65
blood agents/cyanides, 64
chief categories, 63
nerve agents, 63
pulmonary damaging agents, 64
riot-control agents, 64–65
treatment guidelines, 65
vesicants/blister agents, 63–64
risks, 62
sources, 62
terrorism events, 62
Chemistry Information Center for Emergencies (CIQUIME), 227
Chickenpox, smallpox *vs.*, 68
Chlorine, and anhydrous ammonia, 133–134
Choking agents, 77
Chronic aquatic toxicity, 240
CIQUIME, *see* Chemistry Information Center for Emergencies (CIQUIME)
Civil defense sirens, 13
Class 1 Compatibility Group, 151–153
Clothing design, 214
CLP Regulation, 236
CNSS, *see* Committee on National Security Systems (CNSS)
Coalition of the Willing, 169
Command function, 7
Command Staff, 4–5, 89
Committee of Sponsoring Organizations of the Treadway Commission (COSO), 27
Committee on National Security Systems (CNSS), 21–22
Community alarm, *see* Autodialed alarm
Comprehensive Environmental Response, Compensation, and Liability Act (CERCLA), 201
Computer-Assisted Passenger Prescreening System, 29
Coordinator for Counterterrorism, xxvii
Corrosive substances, 233

COSO, *see* Committee of Sponsoring Organizations of the Treadway Commission (COSO)
Covert entry, 41
CTC, *see* DCI Counterterrorist Center (CTC)
Cutaneous anthrax, 69
Cutaneous radiation syndrome, 72
CW, *see* Chemical warfare (CW); Chemical weapons (CW)
Cyber Command, 83
Cyber Shockwave, 83
Cyberwarfare, 81–84
  civil, 84
  denial-of-service attack, 82
  electrical power grid, 82–83
  espionage and national security breaches, 81
  military, 83
  non-profit research, 84
  private sector, 84
  sabotage, 81–82
  terrorism, 83
Cyclotrimethylene trinitramine, 154
Cylinder expansion test, 141
Cylinder fragmentation, 141

DADS, *see* C8-Direct Application Decontamination System (DADS)
DCI Counterterrorist Center (CTC), xxviii
DCS, *see* Distributed control systems (DCS)
Dead-pressing, 143
Decontamination, 181–186
  of chemical warfare and industrial agents, 187–203
    approaches, 196–197
    decontaminants, 187–188
    environmental concerns, 201
    guidelines for mass casualty, 192–196
    methods, 188–192
    overview, 187
    types of chemical victims, 197–201
  definition, 181
  emergency, 185–186
  methods, 182–183
  plan, 184–185
  prevention, 181–182
  for protective clothing reuse, 185
  testing effectiveness of, 183–184
  types, 182
Defense budget, 119
Defense support of civil authorities (DSCA), 49–59
  description, 49–52
  military operations other than war (MOOTW), 52–54
    American deployments, 53
    Australian deployments, 54
    British deployments, 53
    Japanese deployments, 54
Deflagration, 138, 155, 158
Degradation, material, 214
Delta Force, xxvii
Denial-of-service attack (DoS), 82
Department of Defense, 51
Department of Defense Directive 3025.18, 49
Department of Energy (DOE), 164, 177
Department of Homeland Security (DHS), xxvi, 1, 3, 6, 82
DEST, *see* Domestic Emergency Support Team (DEST)
Detonation, 138
  pressure, 141–142
  velocity, 140
DHS, *see* Department of Homeland Security (DHS)
Directorate of Military Support for Domestic Operations (DOMS), 49
Dirty bomb, 74, 169, 175–180
  contact information, 177–178
  effects of RDD, 179
  impact, 175–176
    protective actions, 176
  protection against radiation, 179–180
    potassium iodide tablets, 180
  radiation, 178
  radioactive material
    control, 176–177
    sources, 176–177
    terrorists interest in, 178–179
  RDD and, 178
  risk of cancer, 177
Disposable clothing, 213
Distributed control systems (DCS), 13, 21
Documentation, and rationalization (D&R), 19, 20
DOE, *see* Department of Energy (DOE)
Domestic Emergency Support Team (DEST), xxvi
Domestic terrorism, xxix, 110
DOMS, *see* Directorate of Military Support for Domestic Operations (DOMS)
DoS, *see* Denial-of-service attack (DoS)

Dosimeter devices, 73
D&R, *see* Documentation and rationalization (D&R)
DSCA, *see* Defense support of civil authorities (DSCA)
Duration of exposure, 209
Dynamite, 146

Economic Impacts of Global Terrorism: From Munich to Bali October 2006, 120
Electrical discharge, 141
Electrical explosion, 156
Electrical power grid, 82–83
Electronic eavesdropping, 41
Electronic risk assessment, and vulnerabilities, 43–46
- IT security analysis, 44
- methodology, 44
- tool, 44–46
  - attacks and threats, 45–46
  - risk analysis, 44–45
  - threat analysis, 45

Emergency decontamination, 185–186
Emergency management, 92, 102–103
Emergency Management Institute (EMI), 6
Emergency medical service (EMS), 91–92, 95–96, 100
Emergency Planning and Community Right-to-Know Act (EPCRA), xxvii
Emergency response guidebook (*ERG*), 75–76, 227–245
- boiling liquid expanding vapor explosion (BLEVE), 233–236
  - fires, 235–236
  - mechanism, 233–234
  - water example, 234–235
  - without chemical reactions, 235
- contents, 227–230
  - blue section, 229
  - green section, 229–230
  - orange section, 229
  - white section, 228–229, 230
  - yellow section, 229
- globally harmonized system of classification and labeling of chemicals (GHS), 236–244
  - hazard classification, 236–240
  - hazard communication, 240–244
  - implementation, 244
- shipping list, 230–233
  - classification and labeling summary tables, 231–233
  - contents, 230
  - details, 231
  - recipients, 230–231

EMI, *see* Emergency Management Institute (EMI)
EMS, *see* Emergency medical service (EMS)
Endothermic reaction, 158
Enterprise risk management (ERM), 27–28
Entropic explosives, 157
Environmental hazards, 240
Environmental Protection Agency (EPA), 201
EPA, *see* Environmental Protection Agency (EPA)
EPCRA, *see* Emergency Planning and Community Right-to-Know Act (EPCRA)
Equipment decontamination, 190–192
ERG, see Emergency Response Guidebook (ERG)
ERM, *see* Enterprise risk management (ERM)
Espionage, and national security breaches, 81
Europol, 16–17
Exotic explosive materials, 138–139
Explosive/incendiary weapons of mass destruction, 74
Explosives, 137–159, 231–232
- chemical, 156
- chemical composition, 145–146
  - chemically pure compounds, 145–146
  - mixture of oxidizer and fuel, 146
- classification, 146
  - by composition, 150
  - by physical form, 150
  - by sensitivity, 146–147
  - shipping label classifications, 150–151
  - UNO Hazard Class and Division (HC/D), 151–153
  - by velocity, 147–150
- commercial application, 153
- decomposition, chemical, 138
- deflagration, 138
- detonation, 138
- electrical and magnetic, 156
- exotic, 138–139
- explosive velocity, 154–155
- mechanical and vapor, 156
- natural, 155
  - astronomical, 155

nuclear, 156–157
properties, 157–158
evolution of heat, 157–158
force, 157
fragmentation, 158
initiation of reaction, 158
velocity, 157
properties of explosive materials, 139–145
availability and cost, 139
brisance, 142–143
density, 143
detonation velocity, 140
explosive train, 144
hygroscopicity and water resistance, 143–144
oxygen balance, 145
power, performance, and strength, 141–142
sensitivity determination, 139–140
sensitivity to initiation, 140
stability, 140–141
toxicity, 144
volatility, 143
volume of products, 144–145
Explosive velocity, *see* Detonation, velocity
Exposure *vs.* contamination, 71–72
acute radiation syndrome, 72
cutaneous radiation syndrome, 72
Exterior attack, 41
External security threat, 11
Extremists, 40
Eye damage, 238
Eye irritation, 238

FAA, *see* Federal Aviation Administration (FAA)
False alarms, 14
Fat Man, 161
FBI, *see* Federal Bureau of Investigation (FBI)
Federal Aviation Administration (FAA), xxvi
Federal Bureau of Investigation (FBI), xxviii–xxxii, 171
Federal Emergency Management Agency (FEMA), xxvi–xxvii, 6, 49, 50
Federal Radiological Emergency Response Plan (FRERP), xxvii
Federation of American Scientists, 162
FEMA, *see* Federal Emergency Management Agency (FEMA)
FEST, *see* Foreign Emergency Support Team (FEST)
Field-expedient water decontamination methods, 196–197
Final Presentation, 179
Finance/Administration Section, 90
Fire alarm, 13
Firearms, 42
Fire fighters, 91
Firefighting Resources of California Organized for Potential Emergencies (FIRESCOPE), 2
FIRESCOPE, *see* Firefighting Resources of California Organized for Potential Emergencies (FIRESCOPE)
FIRESCOPE ICS, 2–3
Fire service, 94–95, 99
First Responder, xxi
limitations on certified, xxiii–xxiv
non-traditional, xxv
rescue, xxv
scope of practice, xxiv
skills and limitations, xxiv
traditional, xxv
First responder categories, and capabilities, 90–104
awareness level guidelines, 90–93
emergency management personnel, 92
EMS, 91–92
fire fighters, 91
law enforcement, 90–91
public works employees, 92–93
performance level guidelines, 93–97
EMS, 95–96
fire service, 94–95
hazardous materials, 96–97
law enforcement, 93–94
public works, 97
planning and management level guidelines, 98–104
emergency management, 102–103
EMS, 100
fire service, 99
HAZMAT, 101
law enforcement, 98
public works, 103–104
First responders' resources, 104–106
CBRNE enhanced response force package (CERFP), 106
WMD civil support team (WMD-CST), 104–105

Fission weapons, 162–163
Flammable gas, 232
Flammable liquids, 232, 236
Flammable solids, 232, 237
Force de frappe, 167
Forced entry, 41
Foreign Emergency Support Team (FEST), xxvi
Fragmentation, 158
Fraud management, 17
FRERP, *see* Federal Radiological Emergency Response Plan (FRERP)
Friedrichs Mfg., 15
Fusion weapons, 163–164

Gallois, Pierre Marie, 167
Gallucci, Robert, 168
Gases, 232
Gastrointestinal anthrax, 69
General Staff, 5
Germ cell mutagen, 239
Globally harmonized system of classification and labeling of chemicals (GHS), 236–244
  hazard classification, 236–240
    environmental, 240
    health, 238–239
    of mixtures, 240
    physical, 236–238
  hazard communication, 240–244
    label elements, 241–242
    label format, 242–243
    material safety data sheet, 243–244
    training, 244
  implementation, 244
Gravity bomb, 165
Gregory, Shaun, 171
Gross decontamination, 181, 182
Guerrilla tactics, 111
Guidelines for the Selection of Chemical Protective Clothing, 215

Hand grenades, 42
Harvard University, 169, 170
Hashhashin, 111
Hazard
  classification, 236–240
    environmental, 240
    health, 238–239
    of mixtures, 240
    physical, 236–238
  communication, 240–244
    label elements, 241–242
    label format, 242–243
    material safety data sheet, 243–244
    training, 244
Hazard Class and Division (HC/D), 151
Hazardous material response teams (HMRTs), 186, 203
Hazardous materials (HAZMAT), 95, 96–97, 101, 231
Hazardous materials incidents, xxiii
Hazard pictograms, 241–242
Hazard statements, 242
HAZMAT, *see* Hazardous materials (HAZMAT)
HC/D, *see* Hazard Class and Division (HC/D)
Health hazards, 238–239
Heart rate, 225
Heats of formations, 158
Heat stress, 224–225
  oral temperature, 225
  rate, 225
High explosives, 137, 150
High-risk terrorism, 109
HMRTs, *see* Hazardous material response teams (HMRTs)
Hoax, 17
Hoffman, Bruce, 109
Holy Quran, 114
Homeland Security Presidential Directive-5 (HSPD-5), 1
HSPD-5, *see* Homeland Security Presidential Directive-5 (HSPD-5)
Huang, Reyko, 120
Hydrogen bombs, 163
Hydrogen cyanide, 64
Hygroscopicity and water resistance, 143–144

IAEA, *see* International Atomic Energy Agency (IAEA)
IAP, *see* Incident Action Plan (IAP)
IC, *see* Incident Commander (IC)
ICBMs, *see* Intercontinental ballistic missiles (ICBMs)
ICS, *see* Incident Command System (ICS)
ICS 120-1, *see* Incident Command System Operational System Description (ICS 120-1)
ICS Command, 4
IEDs, *see* Improvised explosive devices (IEDs)
Illicit Nuclear Trafficking Database, 170

Improvised explosive devices (IEDs), 42, 117
Improvised nuclear device, 74
Impulse *vs.* scaled distance, 142
Incendiary devices, 42
Incident Action Plan (IAP), 5, 8
Incident Commander (IC), 4, 88, 89
Incident Command System (ICS), 86–90
command functions/sections, 89–90
command staff, 89
history, 2–3
management, 4–6
modular expansion, 87–88
NIMS and, 3–4
organization, 87
recommended span of control, 90
training, 6–10
Incident Command System Operational System Description (ICS 120-1), 3
Incident locations, and facilities, 8
Individual decontamination, 189–190
Industrial chemicals, as WMD, 131–135
agents and terrorism, 134
chlorine and anhydrous ammonia, 133–134
indicators of chemical incident, 134–135
pesticides and rodenticides, 132–133
tactics, 131–132
Industrial/household incidents, xxii
Infinite-diameter detonation velocity, 142
Information and intelligence management, 8
Information security, 22
Information Security Management System (ISMS), 25
Information Systems Audit and Control Association (ISACA), 28
Information technology (IT) risk, 21–22, 28
Committee on National Security Systems, 21–22
measuring, 22–27
estimation of business impact, 25–27
estimation of likelihood, 23–24
estimation of technical impact, 24–25
occurrence of particular set of circumstances, 22–23
Inhalational anthrax, 69
Insider compromise, 41
Inside Terrorism, 109
Integrated communications, 8
Interagency Intelligence Committee on Terrorism, xxviii
Intercontinental ballistic missiles (ICBMs), 165
Internal security threat, 12
International Atomic Energy Agency (IAEA), 170, 172, 177
International Court of Justice, 167
International terrorism, xxix, 108, 110, 112
Internet, and psychological warfare, 117–118
ISACA, *see* Information Systems Audit and Control Association (ISACA)
Islam
and Holy Quran, 113–114
radical, 114–117
teaching about Jihad and fighting for cause of Allah, 113
ISMS, *see* Information Security Management System (ISMS)
IT, *see* Information technology (IT) risk

Japanese deployments, 54
Joint Operation Center (JOC), 49

Kaspersky, Eugene, 83
Keycard lock, 14
Khinsagov, Oleg, 171
KI, *see* Potassium iodide (KI) tablets
Koran and Bible, violence in, 112–113
Kurtz, George, 84

Law enforcement, 90–91, 93–94, 98
Law enforcement agency (LEA), 16
LEA, *see* Law enforcement agency (LEA)
Liaison Officer, 89
Likelihood estimation, 23–24
Liquid penetration data, 218
Liquid splash-protective suits, 210
Little Boy, 161
Litvinenko, Alexander, 171
Logistics Section, 89–90
Loss prevention, 11
Low explosives, 137, 147, 149

MACP, *see* Military aid to the civil power (MACP)
MACS, *see* Multiagency Coordination System (MACS)
Magnetic explosion, 156
Mail bombs, 41
Management, by objectives, 7
Man-made nuclear weapon, 156
Mass casualty decontamination, 192–196
methods, 194–195

procedures, 195–196
processing tag system, 200
purposes, 193–194
relationships and process, 192–193
Mass Casualty Decontamination Research Team (MCDRT), 195
MCDRT, *see* Mass Casualty Decontamination Research Team (MCDRT)
Mechanical explosion, 156
Medical Management of Chemical Casualties Handbook, 193
Militant groups, 169–170
Military aid to the civil power (MACP), 51
Modern terrorism, 111
Modular organization, 7
Mohammed, Khalid Sheikh, 169
MOOTW, *see* Military operations other than war (MOOTW)
Moving-vehicle bomb, 40
Mueller, Robert S., III, xxviii
Muhammad, Prophet, 114
Multiagency Coordination System (MACS), 2
Multiple-alarm fire, 13
Muscarinic effects, 63
Muslim terrorism, 112

Nanyang Technological University, 15
National Commission on Terrorist Attacks upon the United States, 1
National Contingency Plan (NCP), 201
National Fire Academy (NFA), 6
National Fire Protection Administration (NFPA), 194, 210, 213
National Fire Protection Association, xxv
National Guard CERFP, 106
National Incident Management System (NIMS), 86
and ICS, 3–4, 84–85
National Information Assurance Training and Education Center, 22, 27
National Integration Center (NIC), 85–86
National Interagency Incident Management System (NIIMS), 3
National Preparedness System, 84, 85
National Response Center, xxvii
National Security Telecommunications and Information Systems Security Instruction (NSTISSI), 22
National Source Tracking System, 177
National Wildfire Coordinating Group (NWCG), 2, 3
NATO standardization agreements (STANAGs), 52
Natural disasters, xxii
Natural explosions, 155
NCP, *see* National Contingency Plan (NCP)
Neptunium-237, 162
NERC, *see* North American Electric Reliability Corporation (NERC)
Nerve agents, 63, 78
Neutron bomb, 164
The New York Times, 82, 168
NFA, *see* National Fire Academy (NFA)
NFPA, *see* National Fire Protection Administration (NFPA)
NFPA 471, Recommended Practices for Response to Hazardous Materials Incidents, 194
Night-watchman, 15
NIIMS, *see* National Interagency Incident Management System (NIIMS)
NIMS, *see* National Incident Management System (NIMS)
NIMS Integration Center, 6
Nitroglycerin, 137
Nobel, Alfred, 156
Non-Ambulatory casualties, 198, 200
Non-aqueous methods, 197
Nonflammable gases, 232
Non-state terrorism, 111–112
North American Electric Reliability Corporation (NERC), 82
North Caucasus terrorists, 170
NRC, *see* Nuclear Regulatory Commission (NRC)
NSTISSI, *see* National Security Telecommunications and Information Systems Security Instruction (NSTISSI)
Nuclear agents, 42
Nuclear explosions, 156–157
Nuclear Non-Proliferation Treaty, 162
Nuclear power plant incident, 73
Nuclear Regulatory Commission (NRC), 176, 177
Nuclear weapons (NW), 129–131
Nuclear weapons and radiation, 161–173
delivery systems, 165–166
fission weapons, 162–163
fusion weapons, 163–164
thermonuclear weapons, 163–164
history, 161–162
other uses, 164–165
terrorism, 168–171

incidents involving, 170–171
militant groups, 169–170
warfare strategy, 166–168
NW, *see* Nuclear weapons (NW)
NWCG, *see* National Wildfire Coordinating Group (NWCG)

Office of Emergency Management, 49
O&M, *see* Operation and maintenance (O&M), alarms in
Open Web Application Security Project (OWASP), 23
Operation and maintenance (O&M), alarms in, 13
Operations Section, 89
Operations security (OPSEC), 30
Oral temperature, 225
Organized criminals, 40
OSHA Technical Manual, 225
OWASP, *see* Open Web Application Security Project (OWASP)
Oxidizing agents, and organic peroxides, 233
Oxygen balance, 145

Padilla, José, 171
Particulate penetration data, 218
Particulate protective suits, 217
Pathogen feed stocks, 127
Patterson, Andrew J., 171
PCA, *see* Posse Comitatus Act (PCA)
Peaceful research reactors, 171
Peacekeeper missile, 166
Pedestrian barricades, 15
Penetration, 214
Permeated Contaminants, 182
Permeation, 214
Personal alarm, 13
Personal responsibility, 8
Pesticides, and rodenticides, 132–133
Physical environment, 207–208
Physical hazards, 236–238
Physical security, 15–16
Plague, 70–71
bubonic, 70
overview, 70
pneumonic, 70
septicemic, 70–71
Planning Section, 89
Plutonium-239, 162
Pneumonic plague, 70
Poisonous gases, 232
Posse Comitatus Act (PCA), 51, 52
Potassium iodide (KI) tablets, 73, 177, 180
PPD, *see* Presidential Policy Directive (PPD)
Precautionary statements, 242
Presidential Policy Directive (PPD)-8, 84
Presidential Decision Directive (PDD) 39, xxvi
Pressure *vs.* scaled distance, 142
Primary explosive, 137, 146–147
Priming compositions, 150
Product identifier, 242
Protestors, 40
Public Information Officer, 89
Public works, 92–93, 97, 103–104
Pulmonary damaging agents, 64
Pyrophoric liquid, 236–237

Quadripartite standardization agreements (QSTAGs), 52
Quick Selection Guide to Chemical Protective Clothing, 215

Radiation toxicity/sickness, *see* Acute radiation syndrome
Radiation *vs.* radioactive material, 71
Radical Islam, 114–117
Radioactive and nuclear weapons of mass destruction, 71
dosimeter devices, 73
exposure *vs.* contamination, 71–72
acute radiation syndrome, 72
cutaneous radiation syndrome, 72
methods of protection, 73
potassium iodide (KI) tablets, 73
radiation *vs.* radioactive material, 71
scenarios, 73–74
dirty bomb, 74
hypothetical, 73
improvised nuclear device, 74
nuclear power plant incident, 73
survey meters, 72
Radioactive material
control, 176–177
sources, 176–177
terrorists interest in, 178–179
Radioactive substances, 233
Radiological dispersal device (RDD), 175, 178, 179
Radiological exposure devices, 170
Radiological weapon (RW), 128–129, 170
RBE, *see* Relative bubble energy (RBE)
RDD, *see* Radiological dispersal device (RDD)

RDX/Hexogen, *see* Cyclotrimethylene trinitramine
Readily combustible solids, 237
Reign of Terror, 111
Relative bubble energy (RBE), 142
Religious terrorism, 108, 112–113
  teaching of Islam and fighting for cause of Allah, 113
  violence in Koran and Bible, 112–113
Rem, 178
Reproductive toxins, 239
Resource management, 8
Resource tracking, 9
Respiratory sensitizer, 239
Reusable clothing, 213
Riot-control agents, 64–65
Risk, 12–13
  acceptance, 13
  avoidance, 12
  chemical protective clothing, 224–225
    body water loss, 225
    heat stress, 224–225
  management, 11, 25–26, 27–28
    methodology, 28
    as part of enterprise risk management, 27–28
  reduction, 12
  spreading, 12
  transfer, 12
Risk analysis, 37–39
  electronic assessment and vulnerabilities, 43–46
    IT security, 44
    methodology, 44
    tool, 44–46
  principles of conducting, 46–48
  re-evaluation, 39
  upgrade recommendations, 37–38
Risk IT framework, 28
Robespierre, Maximilien, 111
ROC-IED ABCA US Army training course 2011, 116
Rocks, and clubs, 42
RW, *see* Radiological weapon (RW)

Sabotage, 81–82
Saboteurs, 40
Safety alarms, 13
Safety data sheet (SDS), 243–244
Safety Officer, 89
Salted bomb, 164
Schmidt, Howard, 83
Scientific American, 73
SCT, *see* Secretariat of Communications and Transportation (SCT)
SDS, *see* Safety data sheet (SDS)
Seal Team Six, xxvii
Secondary explosive, 137, 147
Secretariat of Communications and Transportation (SCT), 227
Secretary of Homeland Security, 1
Section Chiefs, 88
Security Council Resolution 1674, 52
Security guards/officers, 15
Security lighting, 15–16
Security management, 11–31
  alarm, 17–21
    concepts, 17–18
    improvement methods, 19–20
    need for, 19
    steps to, 20–21
  categorization, 29–31
    concepts, 30
    significance, 30–31
  information technology (IT) risk, 21–22
    Committee on National Security Systems, 21–22
    measuring, 22–27
  overview, 11
  perceived *vs.* real, 28–29
  policy implementations, 13–17
    access control, 14
    fraud, 17
    intrusion detection, 13–14
    LEA jurisdiction procedures, 16
    non-executive powers jurisdictional coverage, 16–17
    physical security, 15–16
  risk, 27–28
    methodology, 28
    as part of enterprise risk management, 27–28
  types of security threats, 11–13
    external, 11
    internal, 12
    risk, 12–13
Security policy implementations, 13–17, 30
  access control, 14
  alarm devices, 13–14
  fraud management, 17
  LEA jurisdiction procedures, 16
  non-executive powers jurisdictional coverage, 16–17
  physical security, 15–16
Security theater, 29
Security threats, types of, 11–13

external, 11
internal, 12
risk, 12–13
acceptance, 13
avoidance, 12
reduction, 12
spreading, 12
transfer, 12
Self-heating substances, 237
Sensitivity, of explosive
determination, 139–140
to initiation, 140
September 11 attacks, xxxi, 168
Septicemic plague, 70–71
Shipping label classifications, 150–151
el Shukrijumah, Adnan Gulshair, 171
Sicarii, 111
Signal words, 242
Skin corrosion, 238
Skin irritation, 238
Skin sensitizer, 239
SLBMs, *see* Submarine-launched ballistic missiles (SLBMs)
Smallpox, 67–68
vs. chickenpox, 68
clinical features, 68
history, 67
vaccine, 68
Variola major, 67–68
Smoke detector, 13
Sophisticated criminals, 39–40
Span of control, 8, 9
Specific target organ toxicity (STOT), 239
Spies, 40
Splash suits, 217
STANAGs, *see* NATO standardization agreements (STANAGs)
Standoff weapons attack, 41
Stationary-vehicle bomb, 40
Storage temperature, 141
STOT, *see* Specific target organ toxicity (STOT)
Strategic weapons, 167
Submarine-launched ballistic missiles (SLBMs), 165
Subsonic explosions, 155
Subversives, 40
Sunlight, exposure to, 141
Supersonic explosions, 155
Supplemental information, 242
Supplier identification, 242
Support function protective garments, 210, 212
Surface contaminants, 182
Survey meters, 72

Tactical weapons, 166
Target attractiveness, 36
Target organ/systemic toxicity class (TOST), 239
TC, *see* Transport Canada (TC)
Technical impact, 24–25
TEL, *see* Terrorist exclusion list (TEL)
Telegraph, 169
Teller–Ulam design, 163
Terrorism, 83, 107–121
categories, 109–110
cost, 119–121
definition, 108–109
history of, 111–112
origins of modern, 111
rise of non-state, 111–112
turns international, 112
Islam and Holy Quran, 113–114
official U.S. government definition of, 110–111
radical Islam, 114–117
religious, 112–113
teaching of Islam and fighting for cause of Allah, 113
violence in Koran and Bible, 112–113
security and policies of United States, xxv–xxxii
agencies tasked for counterterrorism, xxvii
assignments for specific agencies, xxvi–xxvii
DCI Counterterrorist Center, xxviii
FBI, xxviii–xxxii
Patriot Act, xxvi
terrorist tactics, 117
improvised explosive devices, 117
incident activities, 118–119
internet and psychological warfare, 117–118
training camps, 118
Terrorism Project, 120
Terrorist
interest in radioactive material, 178–179
tactics, 117
improvised explosive devices, 117
incident activities, xxii–xxiii, 118–119
internet and psychological warfare, 117–118
training camps, 118

Terrorist exclusion list (TEL), xxvi
Tertiary explosive, 137, 147
Thermonuclear weapons, 163–164
Threat; *see also* Vulnerability assessment
assessment, 34–37
computer/network internal and external, 43–46
IT security risk assessment and analysis, 44
methodology, 44
vulnerability assessment tool, 44–46
understanding, 39–43
TIC/TIM, *see* Toxic industrial chemicals and materials (TIC/TIM)
Tocsins, 13
Tornado sirens/air raid sirens, *see* Civil defense sirens
TOST, *see* Target organ/systemic toxicity class (TOST)
Toxic, and infectious substances, 233
Toxic industrial chemicals and materials (TIC/TIM), 74–81
chemical warfare agents, 77–80
advantages/disadvantages, 80–81
blister agents, 78
blood agents, 77–78
choking agents, 77
control zones, 79–80
isolate, 79
lethality relative to chlorine, 78–79
nerve agents, 78
Emergency Response Guidebook (ERG) 2012, 75–76
RAIN, 75
recognition, 75
sources, 76–77
Toxicity, 144
Transportation emergency incidents, xxii
Transportation Security Administration (TSA), xxvi
Transport Canada (TC), 227
Triage, 198
TSA, *see* Transportation Security Administration (TSA)
Tsar Bomba, 164

UC, *see* Unified Command (UC)
UN, *see* United Nations (UN)
Unified Command (UC), 5, 7
United Nations (UN), 52
United Nations Dangerous Goods System, 236
United Nations Organization (UNO) Hazard Class and Division (HC/D), 151–153
Class 1 Compatibility Group, 151–153
explosives warning sign, 151
United States' counterterrorism policy, xxv
United States National Guard, 50, 51
United States Occupational Safety and Health Administration, 236
Unity of command, 7, 8
UNO HC/D, *see* United Nations Organization (UNO) Hazard Class and Division (HC/D)
Unsophisticated criminals, 39
Uranium-235, 162
USCYBERCOM, 83
U.S. Department of Transportation (US DOT), 227, 231
U.S. DOT, *see* US Department of Transportation (US DOT)
U.S. Forest Service, 2
U.S. Northern Command, 50
U.S. State Department Coordinator for Counterterrorism, xxv

Vandals, and activists, 40
Vapor cloud explosion (VCE), 236
Vapor explosions, 156
Vapor-protective suits, 210
Variola major, 67–68
VCE, *see* Vapor cloud explosion (VCE)
Vehicle bombs, 43
Velocity of detonation (VoD), *see* Detonation velocity
Vendor data/recommendations, 215
Vesicants/blister agents, 63–64
Visual surveillance, 41
Vulnerability assessment, 35–36

Waltz, Kenneth, 167
Washington, George, 51
Watchman, 15
Water-based decontamination, 194
Waterborne contamination, 42
Weapon miniaturization, 165
Weapon of mass destruction (WMD), 61, 186, 203, 245; *see also* Chemical, biological, radiological, nuclear and explosive (CBRNE)
biological, 65–71
anthrax, 68–70
attractive to terrorists, 65
bioterrorism, 65

characteristics of attacks, 65
critical biological agents, 66–67
genealogical classification, 65–66
plague, 70–71
smallpox, 67–68
chemical, 62–65
classification, 62–65
risks, 62
sources, 62
terrorism events, 62
cyberwarfare, 81–84
civil, 84
denial-of-service attack, 82
electrical power grid, 82–83
espionage and national security breaches, 81
military, 83
non-profit research, 84
private sector, 84
sabotage, 81–82
terrorism, 83
explosive/incendiary, 74
widely used, 74
first responder categories and capabilities, 90–104
awareness level guidelines, 90–93
performance level guidelines, 93–97
planning and management level guidelines, 98–104
first responders' resources, 104–106
CBRNE enhanced response force package (CERFP), 106
WMD civil support team (WMD-CST), 104–105
Incident Command System (ICS), 86–90
command functions/sections, 89–90
command staff, 89
modular expansion, 87–88
organization, 87
recommended span of control, 90
National Incident Management System (NIMS), 84–85
National Integration Center (NIC), 85–86
radioactive and nuclear, 71–74
dosimeter devices, 73
exposure *vs.* contamination, 71–72
methods of protection, 73
radiation *vs.* radioactive material, 71
scenarios, 73–74
survey meters, 72
toxic industrial chemicals and materials (TIC/TIM), 74–81
advantages/disadvantages, 80–81
chemical warfare agents, 77–80
Emergency Response Guidebook (ERG) 2012, 75–76
RAIN, 75
recognition, 75
sources, 76–77
Wilkinson, Paul, 121
WMD, *see* Weapon of mass destruction (WMD)
WMD civil support team (WMD-CST), 104–105